Textile Industry and the Environment - Challenges, Recent Development and Future

Edited by Chi-leung Hui

Published in London, United Kingdom

Textile Industry and the Environment - Challenges, Recent Development and Future
http://dx.doi.org/10.5772/intechopen.1004482
Edited by Chi-leung Hui

Contributors
Anupam Kumar, Carmen Ka Man Chan, Chi-leung Hui, Chi Wai Kan, Fei Bin, James Kar Hei Fang, Jyoti Rani, Mohammad Abul Hasan Shibly, Mohammad Mohsin Ul Hoque, Muhammad Mirza Nouman Baig, Ramratan Guru, Raul Fangueiro, Sakthivel Santhanam, Satyanarayan Panigrahi

First published in London, United Kingdom, 2025 by IntechOpen
IntechOpen is the global imprint of INTECHOPEN LIMITED, registered in England and Wales, registration number: 11086078, 167-169 Great Portland Street, London, W1W 5PF, United Kingdom

For EU product safety concerns: IN TECH d.o.o., Prolaz Marije Krucifikse Kozulić 3, 51000 Rijeka, Croatia, info@intechopen.com or visit our website at intechopen.com.

British Library Cataloguing-in-Publication Data
A catalogue record for this book is available from the British Library

Textile Industry and the Environment - Challenges, Recent Development and Future
Edited by Chi-leung Hui
p. cm.
Print ISBN 978-1-83634-062-1
Online ISBN 978-1-83634-061-4
eBook (PDF) ISBN 978-1-83634-063-8

Meet the editor

Dr. Chi-leung Hui was an associate professor at the Institute of Textiles and Clothing at the Hong Kong Polytechnic University from 1993 to 2021. He is a chartered engineer and is an expert in applying smart systems such as AI in the healthcare industry, finite element modelling in biomedical engineering, sales forecasting in fashion retailing, and service learning in textile and clothing education. He has published over 60 high-impact research papers and has been granted numerous funding for supporting his research work. Currently, Dr. Hui is working on the "AI-enabled in education" project as a Senior Project Fellow under the Department of Computing at Hong Kong Polytechnic University.

Contents

Preface

This book addresses the textile industry's environmental challenges, recent developments, and future directions. We used different research development results for the textile industry to demonstrate how to reach a green environment in future. There are five chapters presented in this book that help achieve such objectives.

Chapter one introduces the textile industry's environmental challenges, recent actions, and future prospects. This chapter outlines the industry's background and its key environmental challenges. Chapter two addresses some of the key sustainability challenges, including environmental, social, and economic challenges encountered by the textile industry. Chapter three explores advanced technology, such as cutting-edge technologies, for reducing the environmental impact caused by dyeing and finishing processes. Chapter four presents the new approach to using AI technology to enhance the accuracy and efficiency of textile production processes. The last chapter gives insight into new alternatives to deal with the challenges caused by the textile industry. All chapters summarise environmental challenges, the recent development and possible solutions, and the future of avoiding such challenges in the textile industry. It is useful for industrial practitioners to design and develop more research on green textiles.

I acknowledge the support that the Publishing Process Manager Mateo Belir and his editorial team provided while editing and processing the book's publication. His support is essential for delivering high-quality book chapters to readers.

Dr. Chi-leung Hui
The Hong Kong Polytechnic University,
Kowloon, Hong Kong

Chapter 1

Introductory Chapter: Textile Industry and the Environment – Challenges, Recent Development and Future

Chi-leung Hui

1. Introduction

1.1 Green textiles – challenges, recent development and future

The textile industry is one of the oldest and key industries in the world. It plays an important role in the global economy. It covers various sub-industries including spanning, weaving, knitting, dyeing and finishing, garment manufacturing, packaging and logistics. The textile industry operates on three major principles including designing, producing, and distributing a wide range of finished and semi-finished products such as bedding, clothing, apparel, medical, and other accessories. The industry is greatly diverse, from traditional handmaking to computer-aided manufacturing and automation. The major countries contributing to the deliverables of the industry include China, India, Bangladesh, Vietnam and European Union (EU) countries. The global market value of the textile industry in 2023 is around USD 1,837.27 billion in 2023 and is anticipated to grow at a compound annual growth rate (CAGR) of 7.4% in revenue from 2024 to 2030 [1]. Although there has been rapid growth and expansion over the past few years, the textile industry still faces challenges in environmental impact, and it has become one of the most polluting industries in the world. The environmental impact includes *water pollution, carbon dioxide emissions, release of harmful chemicals and dyes into the environment, and increasing garment waste under a shorter product life cycle* [2]. In water pollution, the textile industry is a major contributor, especially in cotton treatment and dyeing processes. Water consumption of producing one cotton shirt is around 3,000 liters. Meanwhile, water wastage from dyeing and finishing processes generally contains toxic chemicals which are harmful to humans. Harmful chemicals generated by industry such as dyes, bleaches, and finishing agents could contaminate the environment such as soil and water. The industry would produce natural and man-made fibers like nylon and polyester. Such synthetic materials could release microplastics which pollute waterpipes during washing. In addition, the textile industry produces 8-10% of global carbon emissions from energy-intensive textile manufacturing process and transportation in the global supply chain [2]. Especially for the production of synthetic fibers like polyester, the process would generate further exacerbating emissions that are derived from fossil

IntechOpen

fuels. Furthermore, more consumers like fast fashion, which has generated a culture of increasing textile waste under a shorter product life cycle. Such impacts led the different governments to set up rules and regulations for environmental protection against textile waste and pollution.

2. Green textiles: Challenges

In recent years, consumers, governments, and environmental protection agencies have raised their concerns on how to deal with such challenges and set up strategies for the environmental protection caused by the textile industry. Some governments are implementing stricter regulations on textile waste, chemical pollution, and carbon emissions [3]. In the European Union (EU), they set up a circular economy action plan to promote recycling, reuse, and sustainable design in the textile industry. EU also launches REACH regulations to restrict the use of hazardous chemicals in textile production as well as EU strategy for sustainable and circular textiles to produce textile products more durable, recyclable, and free of hazardous by 2030. In the United States, the federal government encourages textile producers to take responsibility for end-of-life management of textile products. They form a multi-stakeholder initiative for promoting sustainable practices in the textile industry. In China, the government set up green manufacturing standards for textile factories to reduce pollution and improve energy efficiency as well as a circular economy promotion law for encouraging recycling and waste reduction in the textile industry. In India, zero liquid discharge regulation is mandatory for the textile industry how to treat and recycle wastewater at zero discharge level. They also encourage the use of organic cotton and natural dyes for sustainable textile and fashion products. In Bangladesh, there are the highest numbers of certified green textile and garment factories for reducing energy and water consumption as well as stricter enforcement of environmental laws for textile manufacturers. Besides such environmental protection measures by various countries, the textile industry promotes the use of sustainable materials such as organic natural fibers and recycling synthetic fibers which aims to reach net-zero emissions by 2050. Most fashion brands are moving toward the adoption of green practices and the use of reusable materials for textile and apparel products [4]. However, in many developing countries, environmental regulations are not strictly enforced, and the capital for investment of sustainable materials is very huge because advanced technologies such as bio-tech and waterless dyeing are expensive for development and implementation.

In the consumer's viewpoints, they are increasingly in favor of slow fashion, which focuses on quality and durability. The rental and second-hand markets are growing, which are driven by environmentally aware consumers, especially younger ones [5]. However, such slow fashion and second-hand markets are only applicable to younger consumers, not to mature consumers. Mature consumers like to change garment styles for various occasions and seasons. They prefer to buy different colors and styles of garments for daily dressing. Mature consumers have higher purchasing power than young consumers when buying new textile and fashion products. It leads many textile manufacturers to produce various textile and fashion products with different colors and styles to meet their needs. It will increase textile waste and water consumption. Education about environmental issues toward consumers, especially for mature consumers, producers, and the government will have a significant impact on the textile industry.

3. Recent development and future

In future, there are several issues to be resolved for environmental protection toward textile industry. The first issue is to increase consumer awareness of sustainable textile products, especially for mature consumers. For example, it is possible to educate mature consumers to buy only what they need and the desired products are able to be recycled and sustainable. The second issue is to increase transparency and environmental labeling for all consumers. It is better to provide all consumers with the environmental footprint of their textile and fashion products such as carbon dioxide emissions or water consumption in the product label. For example, EU has provided an EU ecolabel for textile and clothing products. Finally, washing and drying of textile production process contributes to a large impact on environment, and the textile producers could provide consumers with better information on how to reduce their environmental impact *via* QR code imprinted on care label, which provides repair and care guidelines when mobile device scanned QR code to connect the textile and fashion product's website.

In conclusion, the textile industry is at a crossroads on balancing the need for innovation and growth to address environmental, social, and economic challenges faced by the industry. While challenges still alive, technological advancements and market demand are driving the textile industry toward more sustainable products to provide large environmental protection. By examining such chapters in this book, the stakeholders in the industry will be well-positioned on how to leverage technology for thriving in the dynamic changes.

Author details

Chi-leung Hui
The Hong Kong Polytechnic University, Kowloon, Hong Kong

*Address all correspondence to: patrick.hui@connect.polyu.hk

References

[1] Textile Market Size, Share & Trends Analysis Report By Raw Material (Wool, Chemical, Silk), By Product (Natural Fibers, Polyester), By Application, By Region, and Segment Forecasts. San Francisco, CA, United States: Grand View Research, Inc.; 2024 – 2030. Market Analysis Report, Report ID: GVR-1-68038-736-0

[2] Naqvi SLH, Nadeem M, Ayub F. Social and environmental impacts in textile production. In: Dye Pollution from Textile Industry. Singapore: Springer Nature; 2024. ISBN: 978-981-97-5340-6

[3] Bibi S, Khan A, Fubing X, Jianfeng H, Hussain S, Khan AN. Impact of environmental policies, regulations, technologies, and renewable energy on environmental sustainability in China's textile and fashion industry. Frontiers in Environmental Science. 2024;**12**:1496454. DOI: 10.3389/fenvs.2024.1496454

[4] Tabibi A. Sustainable fashion: from fast fashion to responsible recycling. 30 Jan 2024. Available from: https://green.org/2024/01/30/sustainable-fashion-from-fast-fashion-to-responsible-recycling/

[5] Nelen J, Voets A. Renting or buying second-hand clothing as a sustainable option. Netherland: Retail and Consumer - Insights and publications - PwC; 2020. Available from: https://www.pwc.nl/en/insights-and-publications/services-and-industries/retail-and-consumer-goods/renting-or-buying-second-hand-clothing-as-a-sustainable-option.html

Chapter 2

Sustainability Challenges of the Textile Industry

Mohammad Abul Hasan Shibly
and Mohammad Mohsin Ul Hoque

Abstract

The textile industry is a significant economic sector with profound environmental, social, and economic challenges. Its high water consumption, pollution, and waste generation contribute to environmental degradation, while labor exploitation and supply chain opacity raise serious social concerns. Fast fashion exacerbates these issues by promoting disposable clothing, leading to excessive resource consumption and waste accumulation. Addressing these challenges requires a multidimensional approach, including technological advancements, regulatory frameworks, and shifts in consumer behavior. Innovations in water-efficient dyeing, chemical-free textile processing, and sustainable material adoption, such as organic cotton and recycled polyester, are key to mitigating environmental impacts. Strengthening labor standards, ensuring supply chain transparency, and promoting fair trade practices are essential for improving social conditions. Additionally, implementing circular economy principles, such as textile recycling and waste reduction initiatives, can enhance sustainability. Regulatory measures, industry collaborations, and consumer awareness play vital roles in driving systemic change. Governments must enforce stricter environmental and labor policies, while brands and manufacturers should prioritize ethical sourcing and sustainable production. The transition to a sustainable textile industry is imperative for long-term economic viability, environmental conservation, and social well-being.

Keywords: sustainability, textile industry, pollution, fast fashion, circular economy, environmental regulations

1. Introduction

The textile industry is one of the most important economic sectors, both regarding its direct contribution in terms of production, revenues, and employment, and the vast downstream and upstream activities. Environmentally, the industry's output has significant impacts: on the one hand, the exploitation of natural resources, particularly water, and on the other hand, the production of waste and mainly wastewater.

In the first case, the industry is one of the most water consumptive international. In fact, according to the Water Footprint Network: "it takes around 2,700 liters of water to produce one cotton shirt. This is enough water for one person to drink for 2-and-a-half years! The water is mainly used to grow the cotton and dye the material" [1].

In addition, the textile industry is a great contributor to environmental pollution. Around 20% of the global wastewater volume comes from the textile sector. The mentioned wastewater is usually a blend of different toxic materials from the dying steps that in the absence of proper treatment will result in losing the negative influence of the aquatic systems and the contamination of the drinking water source [2].

Socially, the textile industry suffers from many criticisms of labor practices. The most striking example is the one that occurred in Bangladesh in 2013. Then, the Rana Plaza building collapsed and cost the lives of over 1100 workers [3]. These labor practices are criticized for the child's or slaves' labor conditions, and the overall working and living conditions, such as the incapacity to have a decent wage and living payment.

From a sociological perspective, the issue is tangled with complications due to the complex nature of the supply chain and the widespread lack of transparency within it. Often, brands and consumers do not know where and how their specific clothing items are made. This lack of transparency makes it challenging to hold manufacturers accountable for continuing abusive practices hidden from the public eye [4].

Economically, the issues are perpetuated by the overall growth of fast fashion, which normalized and promoted the practice of treating clothes as disposable commodities. According to the United Nations Environment Program, annually, the world buys approximately 62 million tons of apparel. However, relatively few people buy only that, which means this demand is not recycled or reused [5]. This leads to the waste of resources as well as the filling-up of landfills with nonbiodegradable waste, leading to systemic problems in the environment. These challenges require a multidimensional solution that would include new technologies, legal regulations, and changes in consumer habits. It is essential for the long-term success of the industry that all those involved in production, from the farmers and day laborers to the CEOs and buyers, take part in these solutions. Ways to address textile sustainability challenges should be pursued in a variety of ways, but the most problematic issues include water consumption, pollution, and waste disposal.

2. Environmental challenges

2.1 Water consumption and pollution

The textile industry is a major water-consuming sector of the economy, particularly as it pertains to natural fibers such as cotton. Cotton growing is known to be one of the thirstiest processes around. To produce a single cotton shirt and trousers, a person uses more than 20,000 liters of water, or 20 cubic meters per kilogram [6]. In the whole world, there are less than several rivers containing this volume at any given time. The key world manufacturers of cotton—India, China, and Pakistan—are coincidentally arid countries and suffer from water stress not merely in terms of ecological balance but also because their populations go thirsty and lack water for sanitation. Moreover, cotton companies cause severe water pollution. Chemicals used in the manufacture of dyes, bleaches, and fixing agents, and a significant proportion of them are poisonous and consist mainly of heavy metals. These substances ruin not just the marine life but also the river habitat. Azo dyes that are used for turbid coloring are known to release cancer-like amines [7]. Furthermore, the environmental and human health dangers of the discharge of untreated or partially treated textiles wastewater caused by eutrophication, the decline in biodiversity, and aquatic ecosystems' destruction have been shown. For instance, the Institute of Public and Environmental Affairs reported that a large proportion of textile factories in China—one of the biggest world producers—concealed their dirty water into the Yangtze and Pearl waterways.

2.2 Challenges

2.2.1 Technological innovations

There is no use pretending that anything is splendid. However, the good news is that there are several ways to implement technological enhancement in the wastewater treatment sphere. First of all, membrane infusion, ozone treatment, and advanced oxidation techniques can be employed to cleanse the wastewater of hazardous substances [8].

2.2.2 Water-efficient practices

Practices that factories can adopt to be more water and energy-efficient and reduce pollution are critical. For example, using a low-liquor ratio dyeing practice can save up to 50% of water compared to regular industrial practices [9]. According to Smith, another way to minimize water consumption and the impact on the environment is by using less water-intensive crops that necessitate fewer chemicals such as organic cotton, or a variety of textile fibers, such as hemp and bamboo.

2.2.3 Regulatory and policy measures

To combat excessive water consumption and water pollution in textile industries, there is a need for more stringent regulatory frameworks and adequate enforcement aimed at ensuring that textile industries adhere to the necessary environmental standards. Policies, as well as appropriate regulatory systems, which offer incentives to platforms or companies that adopt cleaner and more efficient technologies, will also greatly play important roles in transforming the industry. Water stewardship initiatives are of vital importance, since they give emphasis to proper use of water in an industry or at an organizational level.

The challenge of excessive water consumption and pollution of water in the textile industry is big but not too big to the extent that cannot be addressed. The development of modern technologies as well as a concerted effort on the establishment of suitable regulatory frameworks, and the practice of sustainable practices can go a long way to effectively address the problem. This not only sustainability of the textile industry but also ensures that water is conserved effectively for the use of future generations.

2.2.4 Energy consumption

The processes involved in the production of textiles always tend to be energy-intensive since the operation of different machinery for the spinning, weaving, and knitting companies requires very high temperatures for clothes dyeing, printing, and finishing. Notably, the IEA study was conducted and is estimated to make up 10% of the total industrial energy usage [10]. Moreover, it is impossible to produce any textiles without using energy. This means that there will be both economic and environmental consequences due to such usage.

2.2.5 Greenhouse gas emissions

The solar heat trapped by clothes produced from natural materials like cotton and linen in the textile manufacturing processes is responsible for greenhouse gases. Increased heat in the earth's atmosphere causes natural processes and activities like ventilation and cooling that occur when wearing clothes leading to temperature fluctuations. More greenhouse gas emissions are a result of manufacturing that involves the production of clothes and is added to the already existing heating conditions [11].

2.2.6 Energy efficiency measures

Both energy consumption and the greenhouse gas emitted are significantly reduced in industries as a result of better energy efficiency measures. It has been noted that many companies embody poor energy efficiency conditions enhanced by old technology that uses high energies in their production processes. Also, the Ellen MacArthur Foundation reports that the use of coal and natural gas in a textile industry that specializes in making clothes leads to the emission of greenhouse gas emissions (GHG). Conversely, upon scrutinizing the comprehensive procedure of manufacturing and conveying goods across urban and continental expanses, it becomes apparent that this process engenders approximately 1.2 billion tons more CO_2 emissions annually than the aggregate emissions from all international aviation and maritime transport combined [12].

2.2.7 Renewable energy sources

Another effective strategy that may help to mitigate GHG emissions is the transition to renewable energy sources, such as solar, wind, and bioenergy. It is possible to employ these sources of energy directly in the production processes to reduce the dependency on fossil fuels. There are leading textile manufacturers that have already started installing solar panels and wind turbines at their production sites. These companies strive to achieve a considerable reduction in carbon emissions.

2.2.8 Carbon management practices

Implementing comprehensive carbon management practices is essential for monitoring and managing the emissions across the supply chain as well as achieving net-zero emissions. Such practices include regular carbon audits, taking targeted reduction measures based on a sound scientific basis, and engaging in carbon offset projects [13].

2.2.9 Legislative and policy frameworks

Governments and international organizations are very important in the context of the textile industry's energy consumption and emissions profile. Policymakers, by increasing environmental regulations and incentives for clean energy, can influence business people toward choosing more sustainable options. In light of the foregoing, the way items of clothing are manufactured is influenced by policymakers; therefore, it is vital that they are educated and motivated to act in accordance with national or organizational objectives [14].

2.3 Waste generation in the textile industry

The textile industry contributes significantly to the overall amount of waste in its different stages. The waste can be divided into two groups, namely, pre-consumer and post-consumer waste. Since the industry involves a considerable amount of materials and produces even more, every step in the process of textile manufacturing, from fiber production to fabric cutting, generates waste products. The growing popularity of fast fashion among consumers implies that garments are being thrown away at a higher rate.

2.3.1 Pre-consumer waste

This is also referred to as manufacturing waste as it is the waste that is generated from the manufacturing process of cloths. This includes the remnants from the cutting of the fabric that are normally discarded after cutting, yarn waste that is produced as a result of the spinning and weaving and lastly the defective products. For instance, as cited by Allwood et al. another textile manufacturer discards approximately 15–20% of the fabric in the process of making clothing thus wasting both the material and energy [15].

2.3.2 Post-consumer waste

These are the clothes used by consumers and disposed of to the environment as waste. An estimation provided by the Environmental Protection Agency (EPA) government states, the United States produces approximately 16 million tons of post-consumer waste as it disposes of 84% of the waste in landfills and 15% is used to make cloth cushions, among other products. It also disposes 6% of the post-consumer waste by burning it in the incinerator [16].

2.4 Challenges

2.4.1 Recycling and reuse

Reducing waste requires increasing the recycling and reuse of textile materials. Textile recycling involves processing old clothes and textile fibers into new products. However, the recycling rate for textiles is low due to challenges such as technology in processing blended fiber and the limited infrastructure in collection and sorting materials. Closed Loop Partners' Textile Recycling Innovation is an initiative that focuses on overcoming these challenges by supporting new recycling systems and technologies [17].

2.4.2 Circular economy models

The adoption of the circular economy model can effectively reduce waste generation. The model comprises various measures that focus on the lifecycle of materials, including designing for durability, facilitating repair and maintenance, and returning used clothes to the manufacturer for recycling or repurposing. Companies such as Patagonia and H&M have developed take-back programs to persuade consumers to return their used clothes to recycle or resale [18, 19].

2.4.3 Reducing production waste

The generation of waste during production is one of the significant contributors to the waste problems in the textile industry. Technological measures can be implemented to reduce this waste. For instance, computer cutting technology can be used in optimizing fabric remnant layout during cutting. Firms may also adopt the use of a zero-waste design technique whereby cut patterns do not generate any form of scraps since all pieces are used [20].

2.5 Chemical usage in the textile industry

The textile industry is one of the most chemically dependent sectors and utilizes a wide variety of harmful chemicals at multiple stages, including fiber production, fabric treatment, dyeing, and finishing. These harmful chemicals include dyes, bleaches, and finishing agents. The biggest disadvantage to us is that provides the necessary water-repelling properties vitality durability, and other resiliency, at the same time as other necessary attributes to the textiles. Finishing agents react with other chemicals mentioned and alter into something fatal. All of these are very hazardous because this process causes a problem to nature and to human health.

2.5.1 The scope of chemical use

The use of chemicals in textile production is vast. However, a large percentage of the chemicals used in the industry are harmful to both populations' health and the environment. Concerning the environment, azo dyes, for instance, popular for their intense colors, tend to break down to form aromatic amines, which are major carcinogens [21]. Moreover, materials like waterproofing agents and flame-retardants applied to several textile types contain per- and polyfluoroalkyl substances, which are not only persistent but also associated with many health problems. Health impacts also result from untreated or inappropriately treated effluents discharged into water bodies [22].

2.5.2 Environmental and health impacts

A significant face of using toxic chemicals in the textile industry is their discharge into the water bodies. Usually, untreated and ineffectively treated effluents contain these chemicals. Thus, they lead to the occurrence of severe water pollution which immediately impacts drain dwellers, and as a result, these chemicals enter the human food chain [23]. In addition, workers in textile factories are in danger as they are exposed to toxic chemicals leading to different respiratory problems, skin diseases, and other severe health issues.

2.6 Challenges

The textile industry faces significant sustainability challenges that encompass environmental, social, and economic dimensions. Addressing these challenges requires a holistic approach, integrating technological innovations, regulatory measures, and industry-wide commitment to sustainable practices. Below are some of the key challenges affecting the sector:

2.6.1 Environmental degradation

The textile industry is one of the largest contributors to environmental pollution. The extensive use of water, hazardous chemicals, and non-renewable energy sources in textile manufacturing has led to severe ecological consequences. Wastewater discharge containing dyes, heavy metals, and other toxic substances pollutes rivers and groundwater, causing harm to aquatic ecosystems and human health. Furthermore, industry is a major source of greenhouse gas emissions, contributing to global climate change.

2.6.2 Resource overconsumption

The production of textiles is resource-intensive, particularly in terms of water and raw materials. Cotton cultivation, for example, requires significant amounts of water and pesticides, depleting freshwater resources and contributing to soil degradation. Additionally, the over-reliance on synthetic fibers, derived from fossil fuels, raises concerns about sustainability and resource scarcity in the long run.

2.6.3 Waste management and circular economy

A major issue in the textile sector is the large volume of waste generated throughout the production and consumption cycles. The rise of fast fashion has exacerbated this problem, leading to an increase in textile waste that often ends up in landfills or incinerators. Recycling and circular economy initiatives remain underdeveloped, with only a small percentage of textile waste being repurposed or reused. Developing effective waste management strategies, including extended producer responsibility (EPR) programs and closed-loop recycling systems, is crucial for sustainability.

2.6.4 Social and labor concerns

The textile industry has long been criticized for its exploitative labor practices, particularly in developing countries where low wages, poor working conditions, and lack of labor rights protections are prevalent. Child labor and forced labor remain persistent issues in some regions, undermining social sustainability. Ensuring fair wages, safe working environments, and ethical labor practices requires stringent regulations, industry compliance, and consumer awareness.

2.6.5 Supply chain transparency and accountability

The complexity of global textile supply chains poses a significant challenge in ensuring ethical sourcing and environmental responsibility. Many brands struggle to trace the origins of raw materials and the working conditions of their suppliers.

The lack of transparency allows for the continuation of unsustainable and unethical practices. Implementing blockchain technology, certification programs, and stricter regulatory oversight can enhance supply chain visibility and accountability.

2.6.6 Regulatory and compliance challenges

While various environmental and labor regulations exist, enforcement remains inconsistent across different regions. Many textile manufacturers operate in countries with weak regulatory frameworks, allowing for non-compliance with environmental and labor standards. International agreements, such as the Registration, Evaluation, Authorization, and Restriction of Chemicals (REACH) in the European Union, and industry-led initiatives like the Zero Discharge of Hazardous Chemicals (ZDHC) program, aim to standardize sustainable practices, but global adoption remains a challenge.

2.6.7 Consumer awareness and behavioral shifts

Consumer behavior plays a crucial role in shaping the sustainability of the textile industry. The demand for fast fashion, driven by frequent style changes and low-cost production, encourages unsustainable manufacturing practices. Increasing consumer awareness about the environmental and social impact of their purchasing choices and promoting sustainable alternatives, such as eco-friendly textiles and slow fashion, are essential for long-term industry transformation.

2.7 Adoption of greener chemicals

To ameliorate the negative environmental impacts of textile production, all factories using chemicals must use greener products in their processes. Factories should start phasing hazardous chemicals out of their operations and replace them with safer alternatives. ZDHC, which stands for zero discharge of hazardous chemicals, is a foundation that is globally working to promote strategies that lead to zero discharge of hazardous chemicals from the textile and footwear value chain [24].

2.7.1 Improved wastewater treatment

The textile industry also needs to improve wastewater treatment processes. Wastewater resulting from the use of chemicals in the production of textiles must be treated within these facilities to remove all toxins before the water is released into streams or any other running water body. Methods like membrane filtration, advanced oxidation processes, and biological wastewater treatment help eliminate toxic products from water [25].

2.7.2 Regulatory and compliance measures

The enforcement of stricter regulations and compliance with environmental standards are essential for chemical use management in textiles. Governments can strengthen the controls and monitoring, and international agreements like Registration, Evaluation, Authorization, and Restriction of Chemicals (REACH) in the European Union and related certifications may facilitate the regulation to use and dispose of dangerous substances [26].

2.8 Social challenges in the context of the textile industry

Despite being a significant financial driver, the textile industry faces numerous social issues, which may be challenging and potentially dangerous. Labor rights and working conditions, as well as the transparency of the supply chain, remain the primary social concerns that have to be addressed in order to ensure the development of the industry and improve the conditions in which tens of millions of people are working.

2.8.1 Labor conditions

This industry has a long history of enforcing specific rules, which necessitate significant performance improvement. The conditions in many textile factories, particularly in less developed countries, are extremely exploitative. Workers of these shops have to perform long-lasting work under excessive pressure having considerably low wages. Moreover, the accidents that happen also enhance the necessity for significant improvements in labor conditions; in particular, the collapse of Rana Plaza in Bangladesh in April 2013 resulted in over 1175 victims [27].

2.8.2 Child and forced labor

To a particular extent, the same conditions of life can be associated with child and forced labor, which seems a persistent problem in some of the regions where textiles are produced. The practices of these types not only violate human rights but also tend to cause unemployment and poverty. According to the International Labor Organization, over 230 million children and tens of millions of adults are employed in the textile industry where working conditions, in many cases, are characterized as hazardous [28].

2.8.3 Supply chain transparency

The pervasiveness and complexity of the textile supply chain, with numerous stages across the many countries in the world, make transparency very difficult to achieve. The supply chain remains mostly opaque, which can help hide labor abuse and conceal the extent of environmental damage, making it hard to avoid negative practices for both the companies and the consumers. Moreover, the lack of clear data often makes it difficult to attack the problem by holding the brand and manufacturer responsible for the given harmful practice. For these reasons, while the industry has made certain steps toward greater sustainability, the goal remains a difficult one [29].

2.8.4 Gender inequality

While the textile industry employs a great many women, especially in Asia, women often face specific challenges as workers. They work in the presence of discrimination and sexual harassment, with male workers simultaneously much more likely to receive benefits, such as increased pay, job security, access to health insurance, maternity leave, and paid time off. Thus, in order to address these challenges, it appears important to raise the industry's standards on labor and push for adherence to the minimum wage, limits on the workday length, and rigorous health and safety standards for all workers, to be enforced through the buying companies' constant auditing Grayson, 2010. Moreover, the pressure should be put on the manufacturers by refusing to buy from institutions not adhering to these practices [30].

In summary, the addressing of the issue will require a multipronged approach. Both the demands of the manufacturers and brands should be raised, with the demand for transparency and adherence to the environmental and labor standards. At the same time, the consumers should make their small contribution by buying fair trade-certified products.

2.9 Challenges

2.9.1 Improving labor standards

Implementing and enforcing higher labor standards is essential for addressing social challenges in the textile industry. This includes adhering to minimum wage laws, ensuring safe working conditions, and reducing working hours to reasonable limits. International brands have a crucial role to play by demanding compliance from their suppliers and by conducting regular audits to ensure adherence to labor standards [31].

2.9.2 Promoting fair trade

Supporting fair trade practices can also make a significant difference. Fair trade organizations help ensure that workers are paid fair wages and work under safe conditions. By purchasing fair trade-certified products, consumers can contribute to improving the economic well-being of textile workers.

2.9.3 Enhancing transparency and accountability

Increasing transparency within the supply chain is vital for addressing social challenges. Technologies such as blockchain can help track the origin of garments and ensure that they are produced under fair labor conditions. Furthermore, strengthening reporting requirements and supporting whistleblower protections can improve accountability [32].

2.9.4 Supporting worker empowerment

An effective strategy to support worker empowerment is providing them with proper education, training, and opportunities to get engaged in the union. These strategic steps help to increase the bargaining power of the workers and protect their rights. Another solution would be related to the initiatives for gender equality. Such programs imply providing women with the possibility to take corporate governance training to occupy positions where they wish to work. This approach works well with anti-discriminating laws, which are imperative in this sphere [33].

2.10 Potential solutions to the sustainability challenges in the textile industry

The implementation of potential solutions to the textile industry sustainability challenges should be based on a complex approach covering the environmental, social, and economic dimensions. The introduction of innovative initiatives and efforts in cooperation with all supply levels is required to minimize its negative environmental footprint, improve people's working conditions, and make it more sustainable in the long-term perspective.

2.11 Sustainable materials

2.11.1 Adoption of eco-friendly fibers

The improvement of the environmental sustainability of textiles should be associated with the implementation of eco-friendly technologies. An effective solution in this sphere is the adoption of sustainable raw materials. Organic cotton, linen, hemp, and bamboo are the alternatives to traditional cotton, as their production requires fewer pesticides and water [34]. Another approach is related to the application of recycled materials, such as plastic bottles for producing polyester or carpets [15].

2.11.2 Types of eco-friendly fibers

2.11.2.1 Organic cotton

Organic cotton is a type of fiber produced without harmful and toxic materials, including pesticides, fertilizers, and genetically modified organisms. It reduces the amount of water used in cotton production by about 91% compared to traditional cotton. This fiber is less likely to cause soil and water pollution [35].

2.11.2.2 Hemp

Hemp is a type of fiber with a shorter growing cycle and can be planted in areas affected by drought. No fertilizers or pesticides are used in industrial hemp cultivation. Hemp enriches the soil with various nutrients [36].

2.11.2.3 Bamboo

Bamboo is an environmentally friendly option because it grows rapidly and does not require replanting. This plant also absorbs more CO_2 and releases more oxygen than many other types of vegetation. Moreover, when processed appropriately, it can be transformed into a soft and breathable fabric. Nevertheless, the sustainability of bamboo textiles depends on the chemicals used in production processes. Thus, Liu et al. stress the necessity to apply a closed-loop process to recapture and reuse chemicals.

2.11.2.4 Recycled polyester

Similarly, recycled polyester is more sustainable than its virgin counterpart. It is produced from post-consumer plastic bottles and significantly reduces landfill waste. At the same time, it requires less energy to be developed. Another benefit is that already existing plastic is transformed into high-quality fibers, promoting a lower dependency on oil as a major raw material source. On the other hand, the main disadvantage is that this material retains the ability to shed microplastics, which further pollute water reservoirs. Moreover, in the long run, plastic recycling may require replacements of plastic bottles with alternative material returned. In this respect, using recycled polyester clothing might contribute to the growth of waste [37].

2.11.3 Benefits of using eco-friendly fibers

2.11.3.1 Environmental conservation

In general, eco-friendly fibers require less water, fewer chemicals, and lower energy inputs to cultivate or produce them. As a result, there is less environmental degradation and more preservation of vital resources.

2.11.3.2 Reduction of pollutants

Since these materials do not require the use of toxic pesticides and fertilizers, there is a decrease in the number of contaminants washed into local water systems, protecting local wildlife and inhabitants.

2.11.3.3 Supporting biodiversity

Among the various types of sustainable farming practices, the ones that produce eco-friendly fibers also support higher biodiversity levels. For instance, it is stated that organic cotton farms can house up to 30% more insect species than those grown conventionally [38].

2.12 Challenges and future perspectives

Despite the presence of many benefits, using eco-friendly fibers is still challenging, with the relatively high cost of production, limited variety, and quality of materials. It should be agreed that this area requires a more substantial investment in research and development, raising awareness among the population, and more favorable policies for the benefit of all stakeholders.

It is important to note that numerous sustainable fibers and textiles exist today. For instance, cotton is a crop that uses a lot of water, so organic cotton that is grown using more sustainable methods is considered eco-friendly. Moreover, the supply chain for organic cotton is often enhanced, and the use of certifications such as the Global Organic Textile Standard helps consumers make sure the products they buy are eco-friendly. As higher demand can possibly result in lower prices because of economies of scale, such textiles might become more affordable in the future. It should also be noted that the development of new materials is one of the most promising options.

3. Development of new materials

One such material is Piñatex, which is made from pineapple leaves. Mylo is an alternative for leather that is made from mushrooms. It is evident from this example, some new materials are developed from unusual sources, and this approach makes them relatively less resource-intensive [39].

3.1 Emerging sustainable materials

3.1.1 Biodegradable fibers

Emerging sustainable materials are biodegradable fibers that can pose a solution to the issue of textile waste. Traditional synthetic fibers decompose over hundreds of years, but biodegradable fibers naturally decompose in the environment without any negative impact on it. For example, polylactic acid is a biodegradable synthetic fiber that is made from fermented plant starch, typically that of maize. It is compostable and can be used to make fabric that decomposes under industrial composting conditions, reducing the need for landfill space. Lab-grown materials of plant origin also exist. They are developed *via* cell culture, rather than being cultivated. Some examples of such materials are bio-fabricated silk and leather, and Modern Meadow and Bolt Threads are two leading companies in this area [40].

3.1.2 Lab-grown materials

Unlike ordinary leather, lab-grown one demands no animals to be slaughtered, and, thus, emits significantly less greenhouse gases. The water use of the bio-leather and bio-silk production is also reduced. The other promising area is the use of recycled or upcycled materials, for instance, recycled polyester that is made of post-consumer plastic bottles. In summary, these materials use existing waste and, hence, resource use and energy required to make new products is decreased. Upcycled materials are used to make new high-tech items of high value, enabling decreased landfill to use and decreased need for new items due to longer use of the upcycled product [41].

3.2 Recycled and upcycled materials

Recycled materials, such as recycled polyester made from post-consumer plastic bottles, help reduce waste and lower the environmental impact of textile production. These materials use existing waste products, thus conserving raw materials and reducing the energy required for production. Upcycling, on the other hand, involves creatively reusing waste materials to produce new, high-value items, which helps to extend the lifecycle of textiles and reduce landfill waste.

3.2.1 Natural and renewable fibers

Interest in natural fibers like hemp, bamboo, and organic cotton has been increasing, as these fibers offer a rather low environmental and social footprint compared to traditional cotton and many synthetics. Hemp is a resilient and versatile plant that requires minimal water and no pesticides or fertilizers to grow; it also improves the soil. Bamboo is extremely fast-growing and can be processed into soft, breathable fabrics. At the same time, it is essential to diminish their environmental impacts for processing to claim them to be sustainable [42].

3.3 Benefits of working on developing new materials

3.3.1 Environmental conservation

New and innovative materials often have a smaller environmental footprint in comparison with traditional textiles. Furthermore, they are often designed in a way that reduces resource consumption, waste creation, and enhances their biodegradability, thus helping preserve crucial natural resources and protecting ecosystems.

3.3.2 Reduction of greenhouse gas emissions

Innovative technologies like lab-grown leather or recycled polyester can help cut greenhouse gas emissions up to a certain extent, help implement policies, and meet targets set to help address the issue of climate change and invite a more sustainable way of doing business.

3.3.3 Improved material properties

Many new materials also have improved properties: they can be more durable, biodegradable, or comfortable, absorbing, and preventing moisture as required depending on the purpose. Lab-grown, also called *in-vitro*, leather can be tailored to possess certain characteristics spared in traditional leather like a certain strength or flexibility.

3.4 Challenges and future directions

3.4.1 Technical and economic barriers

First of all, working on developing new materials can be challenging technically and bear high initial costs.

3.4.2 Consumer acceptance

Informing consumers about the benefits of new materials and encouraging them to make sustainable choices while buying clothes or other textile products can be essential for their adoption. Transparent labeling and hefty certification fees.

4. Water and energy efficiency

4.1 Advanced dyeing technologies

A good solution would be the implementation of waterless dyeing technologies. For example, CO_2 dyeing uses pressurized carbon dioxide to dye fabric and completely eliminates water usage and wastewater generation.

4.2 Waterless dyeing technologies

Waterless dyeing technologies have become one of the most substantial advancements in reducing water and chemical use in textile processing. Supercritical CO_2 dyeing eliminates water and uses fewer chemicals while ensuring the most resource-efficient and efficient dyeing process [43]. Additionally, the dyeing time is reduced, and the coloration of the fabric is more even. The technique has a relatively high initial investment for equipment but reduces water and chemical consumption and time, so it is also beneficial in an economic dimension [44].

4.3 Digital printing

Another green technology that allows for reducing water usage while maintaining the proper dying of textiles is digital textile printing. This method applies dye through the textile with high accuracy and level of detail, so that the need for water as a diluent is minimized. In screen printing, the use of water-based paste dye was considered more eco-friendly compared to traditional water-insoluble synthetic dye practices. The digital textile printing technique does not require any water, especially for fixation, so it is favorable from an ecological point of view. In addition, digital printing is well-suited for small production series, which helps to avoid excess dye and water usage due to overproduction. Both methods are examples of ways that can be used to reduce excessive water usage in the process of printing and dyeing textiles and, thus, make the textile industry greener [45].

4.4 Natural dyes and bio-based dyeing

Natural dyes, derived from plants, insects, and minerals, are gaining popularity as a sustainable alternative to synthetic dyes. These dyes are biodegradable and less harmful to the environment and human health. Advances in biotechnology have also led to the development of bio-based dyes, produced using microorganisms. These dyes can offer a consistent color palette and reduce dependency on petroleum-based chemicals.

4.5 Energy efficiency in operations

4.5.1 Energy-efficient machinery and process optimization

It is possible to increase energy efficiency through modern equipment and operations that can reduce greenhouse gas emissions and related energy costs. For example, using energy-efficient motors and LED lighting in factories can significantly reduce energy consumption. The use of energy-efficient machinery is another potential approach to reducing energy consumption in the textile manufacturing process. Modern machinery with designated energy-saving features, such as improved motor systems, heat recovery units, and airflow systems, can achieve reductions in overall electricity, fuel, and water requirements within operations [10]. Advanced knitting and weaving equipment also consume less electricity and work at higher speeds, enhancing overall manufacturing productivity with relatively lower energy use levels.

Other possible options to reduce energy consumption and achieve higher efficiency at operations include process optimization. By managing operational aspects more effectively and integrating automation and control processes, it is possible to reduce energy waste. Specifically, batch process optimization through the establishment of more efficient rehanging processes can achieve the desired outcomes. Moreover, the integration of real-time monitoring and predictive maintenance and the reliance on lean manufacturing processes can also ensure low energy consumption in operations. Finally, the integration of renewable energy sources, such as solar, wind, and biomass, in operations will ensure decreased use of fossil fuels and other GHG-producing resources. For example, solar thermal systems can be installed in operations to provide heat for dyeing, drying, and finishing units. As a result, lower GHG emissions and associated costs can be supported, which will also ensure a better reputation of the company as a sustainable products provider [46].

The use of advanced technologies and the implementation of energy-efficient practices bring substantial benefits to the textile production process. Firstly, advanced dyeing technologies and machinery reduce the environmental impact of textile production by lowering water and chemical use and energy consumption and decreasing greenhouse gas emissions. Additionally, these innovations generate cost savings. While the acquisition of new technologies and machinery may require huge initial investments, the long-term operational savings are significant due to decreased water and energy use and reduced waste disposal costs [47]. Moreover, advanced dyeing technologies and energy-efficient machinery enable higher quality as a result of better color adhesiveness and reduced damage to fabrics, as well as increased product flexibility, allowing to meet market demands faster and reduce lead times.

The application of circular economy principles to the textile industry is related to designing out waste and keeping the materials in use. To achieve these objectives, textile production should focus on reducing waste during all stages of the product's life cycle, including the production, use, and end-of-life stages. In the production stage, waste can be minimized by designing textile products in a way that ensures their long-term use, as well as easy repair and recycling. Supply chains should be designed to facilitate post-consumer textile recycling and the development of products made from these recycled materials. At the same time, business models and consumption practices should be transformed to introduce habits like take-back programs, resale or lease services, as well as recycling.

4.5.2 Renewable energy integration

Integrating renewable energy sources, such as solar, wind, and biomass, into textile production can significantly reduce greenhouse gas emissions and reliance on fossil fuels. Solar thermal systems, for instance, can be used to generate the heat required for dyeing and finishing processes. Companies that invest in renewable energy not only benefit from reduced energy costs but also enhance their sustainability credentials [10].

4.6 Benefits of advanced technologies and energy efficiency

4.6.1 Environmental impact reduction

Adopting advanced dyeing technologies and energy-efficient practices significantly reduces the environmental footprint of textile manufacturing. These innovations lower water and chemical use, decrease energy consumption and reduce greenhouse gas emissions.

4.6.2 Cost savings

Although the initial investment in new technologies and energy-efficient machinery can be high, the long-term operational cost savings are considerable. Reduced energy and water consumption, along with lower waste disposal costs, contribute to improved profitability.

4.6.3 Enhanced product quality and flexibility

Advanced dyeing technologies often result in superior product quality, with better color consistency and less fabric damage. Energy-efficient operations can also increase production flexibility, allowing manufacturers to respond more quickly to market demands and reduce lead times.

5. Waste reduction

5.1 Circular economy practices

Promoting a circular economy within the textile industry involves designing out waste and keeping materials in use. Strategies include designing for longevity, facilitating easy repair, and recycling garments at the end of their life cycle [11].

5.2 Key circular economy practices

5.2.1 Design for longevity

Designing textiles and garments for longevity involves creating products that are durable, timeless, and easy to repair. This practice reduces the need for frequent replacements and minimizes waste. High-quality materials and construction techniques are essential for ensuring product durability. Brands like Patagonia are known for their focus on producing long-lasting products and offering repair services to extend the life of their garments [19].

5.2.2 Recycling and upcycling

Recycling is the process of converting used items into new products. Textile recycling can be done mechanically and chemically. The mechanical process turns used fabrics into fibers, which are spun into new yarn. In this way, both natural and synthetic fabrics can be recycled. The chemical process breaks down the synthetic fibers back to their monomers, which can be repolymerized.

Upcycling, on the other hand, is when waste materials, in this case, used clothes and fabrics, are not broken down but instead used for something else. An example of upcycling would be turning old clothes into new fashion pieces or turning industrial textile waste into new products. The upcycling practices add value to end-of-life products and reduce the need for new materials [48].

5.2.3 Take-back schemes and reverse logistics

Take-back schemes are when retailers and manufacturers take-back end-of-life products for recycling or repurposing. The system ensures that a certain percentage of discarded products are returned to the manufacturing cycle and are not simply being thrown away. Reverse logistics is the quick transportation of the end-of-life products from the retailers and customers to the manufacturers. Efficient reverse logistics systems are essential for the success of take-back schemes and recycling initiatives. In the case of textiles, these schemes are particularly important, as the main consumers, fashion companies, do not have the technology or the desire to recycle the items. An example of a take-back scheme is H&M's system, where customers who return their old clothing would get a discount for new purchases [18].

5.2.4 Closed-loop production systems

Closed-loop systems are those where no waste is produced. The "waste" is reused for the production same or new materials. Since the manufacturing technology cannot yet operate at 100% efficiency, the ideal system is a 90–95% efficient one. The difference between the two is the minimal waste that is produced. An example of a closed-loop production system is when the textile can be disassembled, broken down to raw material, and then restricted back to a textile product. Thus, no difference is seen between the original and the processed items, and the same textile could be broken down and remanufactured forever [49].

5.3 Benefits of circular economy practices

5.3.1 Environmental impact reduction

There are several benefits of the circular economy practices. The first is reducing the impact on the environment. The more is recycled and upcycled, the less needs to be made out of new materials. For example, the production of synthetic textiles is extremely resource-consuming and harmful to the environment. Every little bit recycled decreases the pressure on it. The second benefit is economics. Many companies have already established waste disposal systems, and further thinking about their products' end-life and return options to the production cycle does not usually require big additional investment. On the other hand, the image benefits are numerous. A company that is environmentally friendly receives more recognition from the general public.

The circular economy significantly reduces waste generation and resource consumption. The textile industry can lower its environmental footprint by ensuring materials are kept in use for a longer time and then either reused or recycled. This includes reduced landfill waste, as well as saved water and energy and minimized emissions of pollutants. It also allows for new economic opportunities, such as recycling of textiles or jobs in repairing, and remanufacturing. It opens up prospects for new products, materials, and business models.

5.3.2 Economic opportunities

In addition, due to the increased environmental sustainability of products, adopting circular economy processes can help improve a company's reputation and further establish relationships with consumers. Companies begin to provide repair services and take-back their products that are no longer needed by consumers. It is very important to optimize the efficiency of the product life cycle, as well as enhance the consumer's belief in the organization through faithfulness in environmental values. Consumers today are valuing sustainability with an incredible sentiment that businesses should be concerned about if desire to be competitive. Consumer changes in action are not aspects. In addition to significant action taken, the development of the consumer shift in attitude is to simplify the return of portable products for repair through training, education, and the creation of advertisement systems for sustainable demands. However, before being distributed, the requirements should be offered to the consumer. To ensure the greater adoption of the circular economy in business, the active help of policy and business is necessary. In addition to this, it opens up prospects for new economic opportunities such as recycling or jobs in repairing and remanufacturing.

5.3.3 Policy and regulatory support

Government policies and regulations can support the transition to a circular economy by providing incentives for sustainable practices, funding research and development, and enforcing standards for product design and waste management. Collaborative efforts between industry, government, and consumers are necessary to create a supportive environment for circular economy practices.

5.3.4 Textile recycling innovations

Improving textile recycling technologies is important for handling blended fibers and dyes. Chemical recycling is a major method that disassembles the materials at the molecular level and may be used to recycle the fabrics that are not currently recyclable. Mechanical recycling is based on the disassembly of the fabrics into filaments that may be utilized in the production of new garments. The current paper discusses both recycling methods and describes and compares some examples used in different locations [50].

5.4 Mechanical recycling

5.4.1 Fiber-to-fiber recycling

Fiber-to-fiber recycling is a mechanical method that allows the obtaining of fibers to produce new textiles. First, the fabrics are shredded, and the material is divided into fine fibers that are next carded and spun to get yarn. This method is effective for cotton and wool. The common disadvantages of mechanical recycling are that it reduces the length and quality of fibers and is difficult for blended fabrics that are commonly short-fibered. The exciting news concerns developments in separating and reclaiming such materials, such as Texaid Groups' new machine that divides cotton and polyester fibers. As for the advantages, mechanical recycling is cost-efficient, the machinery used is comparatively cheap, and the process is not very labor-intensive. In general, one may conservatively assume that the quality of the material obtained is inferior to the virgin since it is composed not only of new fibers [34].

Other participants in recycling may include rubber cloth and canvas products, rags of various products, and used upholstery. Mechanical recycling is both less sophisticated and the most widespread. Most of the enterprises for processing these materials are placed in developing countries by such organizations as Tex-India, Tong Siang, Nichols, and others, and employ many workers. At the same time, chemical recycling is mainly implemented by larger companies and small examples do not seem to be successful in Russia in this regard.

5.5 Blended fiber recycling

One of the significant challenges in mechanical recycling is dealing with blended fibers, such as polyester-cotton blends. These fabrics are common but difficult to recycle because they require separation of the different types of fibers. Recent advancements include the development of machinery that can separate and recycle blended fibers more effectively. For example, the Texaid Group has developed a process that separates polyester and cotton fibers, enabling the recycling of blended fabrics [49].

5.6 Chemical recycling

5.6.1 Depolymerization

Chemical recycling involves breaking down synthetic fibers, such as polyester and nylon, into their monomers through a process called depolymerization. These monomers can then be repolymerized to create new fibers. This method can produce high-quality recycled fibers that are almost identical to virgin materials. One example is the chemical recycling of polyester through glycolysis or methanolysis, which breaks down the polymer chains and reassembles them.

5.6.2 Solvent-based recycling

Solvent-based recycling processes dissolve textiles in a solvent to separate the different components. This method is particularly useful for mixed-fiber textiles, as it allows for the recovery of both natural and synthetic fibers. The Evrnu technology, for example, uses a solvent-based process to recycle cotton waste into high-quality cellulose fiber, which can be spun into new yarn.

5.7 Emerging technologies

5.7.1 Enzymatic recycling

Enzymatic recycling is an emerging technology that uses enzymes to break down textile fibers into their basic components. This method is environmentally friendly as it operates under mild conditions and uses biodegradable catalysts. Research is ongoing to develop efficient enzymes for recycling different types of fibers, such as cellulose and polyester [51].

5.7.2 Thermal recycling

Thermal recycling processes, such as pyrolysis, convert textile waste into useful chemicals and fuels through high-temperature treatment in the absence of oxygen. This method can handle a wide range of textile waste, including mixed fibers and contaminated materials. The resulting products can be used as raw materials in chemical manufacturing, reducing reliance on fossil fuels [52].

5.8 Benefits of textile recycling innovations

5.8.1 Environmental impact reduction

Textile recycling innovations significantly reduce the amount of waste sent to landfills and incinerators. By transforming waste into valuable resources, these technologies conserve natural resources and decrease greenhouse gas emissions associated with textile production.

5.8.2 Resource efficiency

Recycling textiles conserves water, energy, and raw materials. For example, recycling polyester saves about 30–50% of the energy required to produce virgin polyester. Similarly, recycling cotton can save water and reduce the use of pesticides and fertilizers [19].

5.8.3 Economic opportunities

Innovative recycling technologies create new economic opportunities by generating value from waste. They support the development of a circular economy in the textile industry, fostering new business models and job creation in the recycling and remanufacturing sectors.

5.9 Challenges and future directions

5.9.1 Technological barriers

Despite the progress in recycling technologies, challenges remain, such as improving the efficiency and scalability of these processes. Continued research and development are needed to address these barriers and enhance the quality of recycled fibers.

5.9.2 Consumer participation

Consumer participation is crucial for the success of textile recycling. Educating consumers about the importance of recycling and providing convenient collection systems can increase recycling rates and support the growth of recycling initiatives.

5.9.3 Policy and regulatory support

Government policies and regulations play a vital role in promoting textile recycling. Incentives for recycling, standards for recycled materials, and extended producer responsibility schemes can encourage companies to invest in recycling technologies and adopt sustainable practices.

6. Chemical management

6.1 Safer chemicals and dye processes

Adopting nontoxic dyes and mordants, and reducing the use of harmful chemicals through standards like the OEKO-TEX and Bluesign certifications, helps in minimizing the adverse environmental and health impacts associated with chemical use in textiles [24].

6.2 Challenges of conventional chemicals and dye processes

Conventional textile processing involves numerous hazardous chemicals, including synthetic dyes, fixatives, and softeners. These substances can have severe environmental consequences when discharged into water bodies, leading to water pollution and harm to aquatic life [53]. Additionally, exposure to toxic chemicals poses health risks to workers and nearby communities.

6.3 Safer chemical alternatives

6.3.1 Low-impact dyes

Low-impact dyes are designed to reduce environmental harm. They require less water, generate less waste, and often do not contain heavy metals or toxic substances. These dyes also have higher absorption rates, reducing the amount of dye that washes out into wastewater. Low-impact reactive dyes, for instance, bond more efficiently with fabric fibers, resulting in less dye runoff [54].

6.3.2 Natural dyes

Natural dyes, derived from plant, animal, or mineral sources, offer a sustainable alternative to synthetic dyes. Examples include indigo (from the indigo plant), cochineal (from insects), and madder root. Natural dyes are biodegradable and less harmful to the environment and human health. However, they may require more land and resources to produce and can be less colorfast compared to synthetic dyes.

6.3.3 Bio-based chemicals

Bio-based chemicals, produced from renewable resources, provide an eco-friendly alternative to conventional petroleum-based chemicals. Enzymes, for instance, are increasingly used in textile processing for desizing, bleaching, and finishing fabrics. Enzymatic treatments are effective at lower temperatures and pH levels, reducing energy consumption and chemical use [55].

6.4 Innovative dye processes

6.4.1 Digital printing

Digital printing technology applies dyes directly to fabrics using inkjet printers, significantly reducing water and chemical use. Unlike traditional dyeing methods, which require large amounts of water to rinse excess dye, digital printing uses precise amounts of dye, minimizing waste. This method is suitable for short production runs and custom designs, reducing overproduction and waste [45].

6.4.2 Supercritical CO_2 dyeing

Supercritical CO_2 dyeing is an innovative waterless dyeing process that uses supercritical carbon dioxide as a solvent to dissolve and transport dyes into fabrics. This technology eliminates water use and produces no wastewater. The CO_2 is recaptured and reused in a closed-loop system, making it a highly sustainable dyeing method. While the initial investment is high, the long-term environmental benefits and cost savings are substantial [43].

6.4.3 Plasma technology

Plasma technology uses ionized gas to modify the surface properties of textiles, enabling better dye uptake and reducing the need for water and chemicals. Plasma treatments can also impart functional properties to fabrics, such as water repellency and antimicrobial effects, without the need for additional chemicals. This technology is energy-efficient and environmentally friendly [56].

6.5 Benefits of safer chemicals and innovative dye processes

6.5.1 Environmental protection

Adopting safer chemicals and innovative dye processes significantly reduces water pollution, chemical waste, and greenhouse gas emissions. These practices contribute to cleaner water bodies, healthier ecosystems, and reduced environmental footprints for textile manufacturers.

6.5.2 Health and safety

Using nontoxic, bio-based chemicals and minimizing hazardous waste protects workers and local communities from chemical exposure. Improved safety standards and working conditions enhance the well-being of textile industry employees.

6.5.3 Economic advantages

While the transition to safer chemicals and advanced dyeing technologies may involve initial costs, the long-term economic benefits include reduced waste management expenses, lower energy and water bills, and enhanced product quality. Sustainable practices can also improve brand reputation and consumer loyalty, driving market growth [57].

6.6 Challenges and future directions

6.6.1 Technological and economic barriers

The adoption of safer chemicals and advanced dye processes requires significant investment in research, development, and infrastructure. Small and medium-sized enterprises may face financial challenges in implementing these technologies. Collaborative efforts and financial incentives from governments and industry stakeholders are crucial for overcoming these barriers.

6.6.2 Regulatory support

Strong regulatory frameworks are needed to enforce the use of safer chemicals and sustainable practices in the textile industry. Policies that promote green chemistry, provide subsidies for sustainable technologies, and penalize noncompliance can drive industry-wide change.

7. Labor practices and transparency

7.1 Fair labor initiatives

Ensuring fair wages and safe working conditions through certification schemes like fair trade and SA8000 helps improve the lives of workers in the textile industry [30].

7.2 Key fair labor initiatives

7.2.1 Fair trade certification

Fair Trade certification ensures that textile products are made under fair labor conditions. This certification requires compliance with stringent labor standards, including fair wages, safe working conditions, and the prohibition of child and forced labor. Organizations like Fair Trade USA and the World Fair Trade Organization (WFTO) work to certify textile products and ensure that workers receive fair compensation and benefits [58].

7.2.2 SA8000 Standard

The SA8000 Standard is a leading social certification program that promotes ethical working conditions globally. Developed by Social Accountability International (SAI), SA8000 sets standards for fair wages, working hours, health and safety, and workers' rights to form unions. Companies that comply with SA8000 undergo regular audits to ensure adherence to these standards [59].

7.2.3 Better work program

The better work program is a collaboration between the International Labor Organization (ILO) and the International Finance Corporation (IFC). It aims to improve labor standards and working conditions in the textile industry by providing training and support to factories and workers. The program focuses on issues such as worker safety, gender equality, and fair wages, working with governments, brands, and workers to create sustainable improvements [60].

7.3 Strategies for implementing fair labor initiatives

7.3.1 Worker empowerment

Empowering workers through education and training is crucial for ensuring fair labor practices. Initiatives that provide workers with knowledge about their rights and skills for negotiating better working conditions are essential. Programs that support unionization and collective bargaining can also help workers advocate for fair wages and safe working environments [30].

7.3.2 Supply chain transparency

Increasing transparency in the supply chain helps identify and address labor abuses. Brands can implement traceability systems to monitor their supply chains and ensure compliance with fair labor standards. Technologies like blockchain can enhance transparency by providing an immutable record of transactions and labor practices [32].

7.3.3 Collaboration and partnerships

Collaboration between brands, suppliers, nongovernmental organizations (NGOs), and governments is vital for promoting fair labor practices. Partnerships can facilitate the sharing of best practices, resources, and knowledge, helping to create a more ethical textile industry. Multi-stakeholder initiatives, such as the Ethical Trading Initiative (ETI), bring together various stakeholders to improve labor conditions in global supply chains [61].

7.4 Benefits of fair labor initiatives

7.4.1 Improved working conditions

Fair labor initiatives lead to significant improvements in working conditions, including safer workplaces, reasonable working hours, and fair compensation. These improvements enhance the well-being and productivity of workers, contributing to a more sustainable industry.

7.4.2 Enhanced brand reputation

Brands that commit to fair labor practices can enhance their reputation and build consumer trust. Ethical consumers are increasingly seeking products made under fair labor conditions, and brands that meet these expectations can gain a competitive advantage in the market [50].

7.4.3 Economic benefits

Fair labor practices can lead to economic benefits for both workers and businesses. Workers who receive fair wages and benefits are more motivated and productive, reducing turnover rates and improving overall efficiency. Additionally, ethical practices can attract investment and support from socially responsible investors [62].

7.5 Challenges and future directions

7.5.1 Enforcement and compliance

Ensuring compliance with fair labor standards can be challenging, especially in regions with weak regulatory frameworks. Robust monitoring and enforcement mechanisms are needed to ensure that factories adhere to fair labor practices. Independent audits and worker feedback systems can help identify and address noncompliance.

7.5.2 Cost implications

Implementing fair labor practices can involve higher costs for brands and suppliers. While these costs can be offset by the long-term benefits of a motivated workforce and enhanced brand reputation, initial investments may be a barrier for some companies. Financial incentives and support from governments and international organizations can help mitigate these costs.

8. Regulatory and policy support

8.1 Supportive legislation

Government policies that encourage sustainable practices through tax incentives, subsidies for clean technologies, and stricter environmental regulations can drive industry-wide changes [14].

8.2 Key legislative areas

8.2.1 Environmental protection

Governments can enforce strict environmental regulations to control the pollution levels associated with textile production. This involves:

- *Water and air quality standards*: Regulations such as the Clean Water Act in the U.S. and the EU Water Framework Directive in Europe set limits on pollutants released into water bodies and the atmosphere, ensuring textile factories treat their waste before disposal.

- *Chemical use regulations*: Legislation like the EU's Registration, Evaluation, Authorization, and Restriction of Chemicals (REACH) mandates the safe production and use of chemicals, aiming to improve human health and the environment.

- *Resource efficiency and waste reduction*: Laws that encourage or mandate recycling and sustainable waste management practices. For example, the EU's Waste Framework Directive pushes for increased recycling and reuse within the textile industry, setting ambitious recycling targets for member states.

- *Water pollution control*: The Clean Water Act in the United States sets standards for discharges of pollutants to ensure that surface waters meet water quality standards. Similar regulations are enforced in the European Union under the Water Framework Directive, which aims to achieve good qualitative and quantitative status for all water bodies [63].

- *Chemical management*: The Registration, Evaluation, Authorization and Restriction of Chemicals (REACH) regulation in the European Union requires companies to identify and manage the risks linked to the substances they manufacture and market in the EU [26].

- *Waste management*: The EU Waste Framework Directive mandates measures to prevent or reduce adverse impacts on the environment and human health from the generation and management of waste [64].

8.2.2 Labor standards

Improving labor conditions in textile factories is critical for addressing social sustainability issues. This can be enforced through:

- *Labor rights laws*: These laws ensure fair wages, reasonable working hours, and safe working conditions. The International Labor Organization (ILO) provides guidelines, although the enforcement is up to individual countries.

- *Fair labor standards*: The International Labor Organization's conventions, such as Convention No. 87 (Freedom of Association) and Convention No. 98 (Right to Organize and Collective Bargaining), provide a framework for labor rights that member states are encouraged to adopt and enforce (ILO Conventions, various years).

- *Prohibition of child and forced labor*: Strong legal frameworks are needed to combat child labor and forced labor in the textile industry. The U.S. has implemented regulations like the Child Labor Laws and the Fair Labor Standards Act to protect against these practices [65].

- *Right to organize*: Ensuring that textile workers have the right to form and join trade unions without discrimination or repression is vital for promoting fair labor practices. Legislation supporting union rights and collective bargaining can lead to significant improvements in worker treatment and pay.

- *Health and safety regulations*: Occupational Safety and Health Administration (OSHA) standards in the United States regulate safety and health conditions in most private industries, including textiles [66].

8.2.3 Trade and economic policies

Trade agreements and economic policies can incentivize or mandate sustainable practices:

- *Import/export regulations*: Tariffs and nontariff barriers can be adjusted to favor environmentally friendly and ethically produced textiles.

- *Sustainable trade agreements*: Modern trade agreements, such as the Comprehensive and Progressive Agreement for Trans-Pacific Partnership (CPTPP), often include chapters on environmental and labor standards that signatory countries must adhere to [67].

- *Subsidies and incentives*: Financial incentives for companies that invest in sustainable technology, waste reduction, and ethical labor practices can accelerate the adoption of sustainable practices within the industry (National and regional government publications).

- *Sustainable public procurement policies*: Governments can lead by example by prioritizing the procurement of sustainable textile products for public sector use, thus creating a significant demand for sustainable textiles.

8.2.4 Transparency and reporting requirements

Mandatory reporting on sustainability metrics can improve transparency and accountability in the textile industry.

- *Corporate sustainability reporting*: The EU Non-Financial Reporting Directive (2014/95/EU) requires large companies to disclose certain information on the way they operate and manage social and environmental challenges [68].

- *Supply chain transparency*: The California Transparency in Supply Chains Act requires retailers and manufacturers doing business in the state to disclose their efforts to eradicate slavery and human trafficking from their direct supply chains [69].

8.2.5 Implementation strategies

- *Enforcement mechanisms*: Effective enforcement mechanisms are critical for ensuring compliance with environmental and labor laws. This might include regular inspections, penalties for noncompliance, and support for companies transitioning to sustainable practices.

- *Education and training*: Legislators can fund programs that educate manufacturers and workers about sustainable practices and technologies, thereby facilitating smoother transitions to greener methods.

- *Public-private partnerships*: Collaborations between governments, industry players, and NGOs can foster innovation and spread best practices across the textile industry.

8.2.6 Benefits of supportive legislation

- *Environmental impact*: Reduced pollution and waste production, contributing to global efforts to combat climate change and preserve biodiversity.

- *Social impact*: Improved working conditions and protection of workers' rights, leading to a better quality of life for employees in the textile industry.

- *Economic impact*: Creation of a level playing field where companies compete to innovate more sustainable and efficient production techniques, potentially leading to job creation in new areas of textile recycling and sustainable materials production.

8.2.7 Challenges

- *Global enforcement*: The global nature of the textile supply chain complicates enforcement, requiring international cooperation to ensure that sustainability standards are met worldwide.

- *Economic impacts*: Transitioning to sustainable practices might initially increase operational costs, which could be challenging for small and medium-sized enterprises.

9. Conclusion

The textile industry, as a cornerstone of global economic, cultural, and social frameworks, faces profound sustainability challenges that necessitate immediate and comprehensive responses. These challenges span environmental, social, and economic domains, each intertwined with the others, reflecting the complexity and scale of the industry's impact. Environmentally, the industry's massive consumption of water and contribution to pollution exemplify the urgent need for sustainable practices. The cultivation of cotton and the dyeing and treatment processes are particularly water-intensive and pollutive, threatening both ecological balance and human health. Innovations such as advanced water treatment technologies and low-liquor ratio dyeing are vital in mitigating these impacts but require widespread adoption, supported by robust regulatory frameworks.

Socially, the industry's labor practices often involve exploitative conditions, highlighted by disasters such as the Rana Plaza collapse. The prevalence of child labor forced labor, and gender inequality within the industry's workforce calls for a stronger enforcement of fair labor standards and enhanced transparency across supply chains. Legislative support, coupled with initiatives like fair trade certification and the SA8000 Standard, can significantly improve labor conditions. However, achieving these improvements also depends on the empowerment of workers and the ethical commitments of global brands.

Economically, the drive toward fast fashion exacerbates resource wastage and environmental degradation, with most apparel ending up in landfills or incinerators. This unsustainable economic model highlights the critical need for a shift toward circular economy practices, where material reuse and recycling are prioritized over disposal. Legislative frameworks and policies must incentivize such a shift, promoting technological innovation and infrastructure development for textile recycling and waste management.

To address these multi-faceted challenges, a coordinated effort involving technological innovation, regulatory reform, and shifts in consumer behavior is imperative. The path forward must engage stakeholders at all levels—from policymakers and industry leaders to consumers and workers—each playing a pivotal role in forging a sustainable future for the textile industry. By implementing supportive legislation, enhancing supply chain transparency, and fostering international cooperation, the textile industry can move toward sustainability, ultimately contributing to environmental conservation, social justice, and economic viability for future generations.

Acknowledgements

The author acknowledges the use of quillbot and grammarly for language polishing of the manuscript.

Author details

Mohammad Abul Hasan Shibly* and Mohammad Mohsin Ul Hoque
National Institute of Textile Engineering and Research, Dhaka, Bangladesh

*Address all correspondence to: hasanduet002@gmail.com

References

[1] Hoekstra AY. The Water Footprint of Modern Consumer Society. Routledge; 2019. DOI: 10.4324/9780429424557

[2] Petit B, Giesen T. Environmental Impact of the Textile and Clothing Industry: What Consumers Need to Know [European Parliament Briefing]. European Parliament; 2011. Available from: https://www.europarl.europa.eu/RegData/etudes/BRIE/2019/633143/EPRS_BRI%282019%29633143_EN.pdf

[3] Siegle L. To Die For: Is Fashion Wearing Out the World? HarperCollins; 2013. ISBN 978-0007484214. Available from: https://www.harpercollins.co.uk

[4] Jia F, Seuring S, Chen L, Azadegan A. Guest editorial: Supply chain transparency: Opportunities, challenges, and risks. International Journal of Operations & Production Management. 2024;**44**(9):1525-1538. DOI: 10.1108/IJOPM-09-2024-992

[5] Ellen MacArthur Foundation. A New Textiles Economy: Redesigning Fashion's Future. 2017. Available from: https://www.ellenmacarthurfoundation.org/publications/a-new-textiles-economy

[6] World Wildlife Fund. The Impact of Cotton on Water Resources and Ecosystems [WWF Cotton Report]. World Wildlife Fund; 2013. Available from: https://www.worldwildlife.org/

[7] Water resources contamination and health hazards by textile industry effluent and glance at treatment techniques: A review. Waste Management Bulletin. 2024;**1**(4):158-163. DOI: 10.1016/j.wmb.2023.09.002

[8] Liu Y-c, Fei Y-h, Li Y-s, Bao X-l, Zhang P-w. Pollution source identification methods and remediation technologies of groundwater: A review. China Geology. 2024;**7**(1):125-137. ISSN 2096-5192. DOI: 10.31035/cg2022080

[9] Das R. Application photocatalysis for treatment of industrial waste water—A short review. Open Access Library Journal. 2014;**1**:1-17. DOI: 10.4236/oalib.1100713

[10] Al-Obaidi KM, Hossain M, Alduais NAM, Al-Duais HS, Omrany H, Ghaffarianhoseini A. A review of using IoT for energy efficient buildings and cities: A built environment perspective. Energies. 2022;**15**(16):5991. DOI: 10.3390/en15165991

[11] Plakantonaki S, Kiskira K, Zacharopoulos N, Chronis I, Coelho F, Togiani A, et al. A review of sustainability standards and ecolabeling in the textile industry. Sustainability. 2023;**15**(15):11589. DOI: 10.3390/su151511589

[12] Hasanbeigi A, Price L, Lin E. Emerging energy-efficiency and CO2 emission-reduction technologies for cement and concrete production: A technical review. Renewable and Sustainable Energy Reviews. 2012;**16**(8):6220-6238. ISSN 1364-0321. DOI: 10.1016/j.rser.2012.07.019

[13] United Nations Framework Convention on Climate Change. Fashion Industry Charter for Climate Action [UNFCCC News Release]. United Nations Framework Convention on Climate Change; 2019. Available from: https://unfccc.int/climate-action/sectoral-engagement-for-climate-action/fashion-charter

[14] World Trade Organization. Trade and Environmental Sustainability:

Implications and Strategies for the Textile and Clothing Sector [WTO Discussion Paper]. World Trade Organization; 2020. Available from: https://www.wto.org/english/res_e/publications_e/disc_paper5_e.htm

[15] Allwood JM et al. Well Dressed? The Present and Future Sustainability of Clothing and Textiles in the United Kingdom. University of Cambridge; 2006. Available from: https://www.ifm.eng.cam.ac.uk/uploads/Resources/Other_Reports/UK_textiles.pdf

[16] Environmental Protection Agency. Advancing Sustainable Materials Management: 2017 Fact Sheet. U.S. Environmental Protection Agency; 2019. Available from: https://www.epa.gov/sites/default/files/2019-11/documents/2017_facts_and_figures_fact_sheet_final.pdf

[17] Closed Loop Partners. Innovations in Textile Recycling: The Next Horizon. Closed Loop Partners; 2020. Available from: https://www.closedlooppartners.com/wp-content/uploads/2020/10/CLP_TextileRecyclingReport_2020.pdf

[18] H&M Group. Sustainability Performance Report 2020. H&M Group; 2021. Available from: https://hmgroup.com/wp-content/uploads/2021/03/HM-Group-Sustainability

[19] Blake A. Inside Patagonia's Operation to Keep you from Buying New Gear. The Washington Post; 2018. Available from: https://www.washingtonpost.com/business/inside-patagonias-operation-to-keep-you-from-buying-new-gear/2018/08/31/d3d1fab4-ac8c-11e8-b1da-ff7faa680710_story.html

[20] Rissanen T, McQuillan H. Zero Waste Fashion Design. 2nd ed. Bloomsbury Visual Arts; 2023. ISBN 9781350116979. Available from: https://www.bloomsbury.com/us/zero-waste-fashion-design-9781350116979/

[21] Rovira J, Domingo JL. Human health risks due to exposure to inorganic and organic chemicals from textiles: A review. Environmental Research. 2019;**168**:62-69. ISSN 0013-9351. DOI: 10.1016/j.envres.2018.09.027

[22] Lindstrom AB, Strynar MJ, Libelo EL. Polyfluorinated compounds: Past, present, and future. Environmental Science & Technology. 2011;**45**(19):7954-7961. DOI: 10.1021/es2011622

[23] Jegatheesan V, Pramanik BK, Chen J, Navaratna D, Chang C-Y, Shu L. Treatment of textile wastewater with membrane bioreactor: A critical review. Bioresource Technology. 2016;**204**:202-212. ISSN 0960-8524. DOI: 10.1016/j.biortech.2016.01.006

[24] Zero Discharge of Hazardous Chemicals (ZDHC). ZDHC Guidelines for Wastewater Treatment in Textile and Leather Industries [Version 1.1]. Zero Discharge of Hazardous Chemicals Foundation; 2020. Available from: https://downloads.roadmaptozero.com/output/ZDHC-Wastewater-Guidelines

[25] Schwarzenbach R, Egli T, Hofstetter T, Gunten U, Wehrli B. Global water pollution and human health. Annual Review of Environment and Resources. 2010;**35**. DOI: 10.1146/annurev-environ-100809-125342

[26] European Parliament and Council. Regulation (EC) no 1907/2006 concerning the registration, evaluation, authorisation and restriction of chemicals (REACH). Official Journal of the European Union. 2006;**L 396**:1-849. Available from: https://eur-lex.europa.eu/legal-content/EN/TXT/?uri=CELEX%3A32006R1907

[27] Berik G, Van Der Meulen Rodgers Y. Options for enforcing labor standards: Lessons from Bangladesh and Cambodia. Journal of International Development. 2010;**22**:56-85. DOI: 10.1002/jid.1534

[28] International Labour Organization (ILO). Global Estimates of Child Labour: Results and Trends, 2012-2016. International Labour Organization; 2017. Available from: https://www.ilo.org/publications/global-estimates-child-labour-results-and-trends-2012-2016

[29] Reza N, Du Plessis JJ. The garment industry in Bangladesh, corporate social responsibility of multinational corporations, and the impact of COVID-19. Asian Journal of Law and Society. 2022;**9**(2):255-285. DOI: 10.1017/als.2022.9

[30] Anner M. Binding Power: The Sourcing Squeeze, Workers' Rights, and Building Safety in Bangladesh since Rana Plaza. Free University of Berlin; 2018. Available from: https://www.wiwiss.fu-berlin.de/forschung/Garments/Medien/2018-Anner-Research-Report-Binding-Power.pdf

[31] Gereffi G, Regini M, Sabel C. On Richard M. Locke, the Promise and Limits of Private Power: Promoting Labor Standards in a Global Economy. New York: Cambridge University Press. Socio-Economic Review. 12. 219-235; 2013. DOI: 10.1093/ser/mwt023

[32] Arthur PJ. Mol, transparency and value chain sustainability. Journal of Cleaner Production. 2015;**107**:154-161. ISSN 0959-6526. DOI: 10.1016/j.jclepro.2013.11.012

[33] Barrientos S. Gender production networks: Sustaining cocoa-chocolate sourcing in Ghana and India. SSRN Electronic Journal. 2013. DOI: 10.2139/ssrn.2278193

[34] Boone T. Sustainable fashion and textiles: Design journeys by Kate Fletcher (Earthscan, 2008). Fashion Practice. 2009;**1**(2):271-274. DOI: 10.2752/175693809X469229

[35] Textile Exchange. Organic Cotton Market Report. Textile Exchange; 2017. Available from: https://textileexchange.org/app/uploads/2024/02/Textile-Exchange_Organic-Cotton-Market-Report_2017.pdf

[36] Russo D. Hemp as an Agricultural Commodity. Congressional Research Service Report; 2019. Available from: https://crsreports.congress.gov/product/pdf/RL/RL32725

[37] Abrishami S, Shirali A, Sharples N, Kartal GE, Macintyre L, Doustdar O. Textile recycling and recovery: An eco-friendly perspective on textile and garment industries challenges. Textile Research Journal. 2024;**94**(23-24):2815-2834. DOI: 10.1177/00405175241247806

[38] Mehmeti A, Abdelhafez A, Ellssel P, Todorovic M, Calabrese G. Performance and sustainability of organic and conventional cotton farming systems in Egypt: An environmental and energy assessment. Sustainability. 2024;**16**:6637. DOI: 10.3390/su16156637

[39] Patti A, Cicala G, Acierno D. Eco-sustainability of the textile production: Waste recovery and current recycling in the composites world. Polymers. 2020. DOI: 10.3390/polym13010134

[40] Vink ETH, Rábago KR, Glassner DA, Gruber PR. Applications of life cycle assessment to NatureWorks™ polylactide (PLA) production. Polymer Degradation and Stability. 2003;**80**(3):403-419. ISSN 0141-3910. DOI: 10.1016/S0141-3910(02)00372-5

[41] Kim J, Kim Y, Yoo J, et al. Organic batteries for a greener rechargeable world.

Nature Reviews Materials. 2023;**8**:54-70. DOI: 10.1038/s41578-022-00478-1

[42] Joshi M, Ali SW, Rajendran S. Ecofriendly antimicrobial finishing of textiles using bioactive agents based on natural products. Indian Journal of Fibre & Textile Research. 2015;**34**(3):295-304. ISSN 0971-0426 (Print), 0975-1025 (Online). Available from: https://nopr.niscpr.res.in/handle/123456789/6083

[43] Beckman EJ. Supercritical and near-critical CO2 in green chemical synthesis and processing. The Journal of Supercritical Fluids. 2004;**28**(2-3):121-191. ISSN 0896-8446. DOI: 10.1016/S0896-8446(03)00029-9

[44] Amenaghawon A, Anyalewechi C, Kusuma H, Mahfud M. Applications of supercritical carbon dioxide in textile industry. 2020. DOI: 10.1016/B978-0-12-817388-6.00013-1

[45] Kornit Digital. Kornit 2022 Impact Report: Environmental, Social, and Governance Progress [White Paper]. Kornit Digital; 2022. Available from: https://kornit.gcs-web.com/static-files/69258658-f8f1-4c0a-9211-6b5bb3e6febb

[46] Sustainable Apparel Coalition. The Higg Index: Tools for the Apparel and Footwear Industry to Measure and Improve Sustainability. Sustainable Apparel Coalition; 2019. Available from: https://product.higg.org

[47] Worrell E, Galitsky C, Price L. Energy Efficiency Improvement and Cost-Saving Opportunities for the Textile Industry. Environmental Energy Technologies Division, Lawrence Berkeley National Laboratory; 2010. Available from: https://www.osti.gov/servlets/purl/927881

[48] Sung K. A review on upcycling: Current body of literature, knowledge gaps and a way forward. In: Proceedings of the 17th International Conference on Environmental, Cultural, Economic and Social Sustainability. World Academy of Science, Engineering and Technology; 2015. Available from: https://www.waset.org/conference/2015/04/venice/ICECESS

[49] Palm D, Elander M, Gíslason S. Towards a New Nordic Textile Commitment: Collection, Sorting, Reuse and Recycling of Textiles. Nordic Council of Ministers; 2014. Available from: https://www.norden.org/en/publication/towards-new-nordic-textile-commitment

[50] Nielsen. The Evolution of the Sustainability Mindset. Nielsen; 2018. Available from: https://www.nielsen.com/insights/2018/what-sustainability-means-today

[51] Egan J, Wang S, Shen J, Baars O, Moxley G, Salmon S. Enzymatic textile fiber separation for sustainable waste processing. Resources, Environment and Sustainability. 2023;**13**:100118. DOI: 10.1016/j.resenv.2023.100118

[52] Kaur G, Uisan K, Ong KL, Lin CSK. Recent trends in green and sustainable chemistry & waste valorisation: Rethinking plastics in a circular economy. Current Opinion in Green and Sustainable Chemistry. 2018;**9**:30-39. ISSN 2452-2236. DOI: 10.1016/j.cogsc.2017.11.003

[53] Kant R. Textile dyeing industry an environmental hazard. Natural Science. 2012;**4**(1):22-26. DOI: 10.4236/ns.2012.41004

[54] Lewis DM. Coloration technology: Advances and developments. Journal of the Society of Dyers and Colourists. 2014;**130**(4):223-228. DOI: 10.1111/cote.12114

[55] Cavaco-Paulo A, Gübitz GM. Textile Processing with Enzymes. Woodhead Publishing; 2003. ISBN 978-1855736703

[56] Marcandalli B, Riccardi C. Plasma treatments of fibres and textiles. 2007. DOI: 10.1533/9781845692575.2.282

[57] Resta B, Gaiardelli P, Pinto R, Dotti S. Enhancing environmental management in the textile sector: An organisational-life cycle assessment approach. Journal of Cleaner Production. 2016;**135**:620-632. ISSN 0959-6526. DOI: 10.1016/j.jclepro.2016.06.135

[58] Fair Trade USA. Fair Trade Certified Apparel. Fair Trade USA; 2020. Available from: https://www.fairtradecertified.org/

[59] Social Accountability International. SA8000 Standard. Social Accountability International; 2017. Available from: https://sa-intl.org/wp-content/uploads/2020/02/SA8000Standard2014.pdf

[60] International Labour Organization, & International Finance Corporation. Better Factories Cambodia: An Industry and Compliance Review – Annual Report 2018. 35th ed. International Labour Office; 2018. ISSN: 2227-958X. Available from: https://betterwork.org/wp-content/uploads/BFC-Annual-Report-2018-1.pdf

[61] Ethical Trading Initiative (ETI). ETI Annual Report 2019. Ethical Trading Initiative; 2019. Available from: https://www.ethicaltrade.org/insights/resources/eti-annual-review-2018-19

[62] Bair J. Corporations at the United Nations: Echoes of the new international economic order? Humanity: An International Journal of Human Rights, Humanitarianism, and Development. 2015;**6**:159-171. DOI: 10.1353/hum.2015.0001

[63] European Parliament and Council. Directive 2000/60/EC of the European Parliament and of the council establishing a framework for community action in the field of water policy. Official Journal of the European Union. 2000. Available from: https://eur-lex.europa.eu/legal-content/EN/TXT/?uri=CELEX%3A32000L0060

[64] European Parliament and Council. Directive 2008/98/EC of the European Parliament and of the council on waste and repealing certain directives. Official Journal of the European Union. 2008. Available from https://eur-lex.europa.eu/legal-content/EN/TXT/?uri=CELEX%3A32008L0098

[65] U.S. Department of Labor. Fair Labor Standards Act. U.S. Government Printing Office; 1938. Available from: https://fraser.stlouisfed.org/title/fair-labor-standards-act-1938-5567/fulltext

[66] Occupational Safety and Health Administration. Occupational Safety and Health Act of 1970. Public Law 91-596. U.S. Government Printing Office; 1970. Available from: https://www.osha.gov/laws-regs/oshact/completeoshact

[67] Comprehensive and Progressive Agreement for Trans-Pacific Partnership. Comprehensive and Progressive Agreement for Trans-Pacific Partnership. 2018. Available from: https://www.dfat.gov.au/trade/agreements/in-force/cptpp/comprehensive-and-progressive-agreement-for-trans-pacific-partnership

[68] European Parliament and Council. Directive 2014/95/EU of the European Parliament and of the council amending directive 2013/34/EU as regards disclosure of non-financial and diversity information by certain large undertakings and groups. Official

Journal of the European Union. 2014;**L 330**:1-9. Available from: https://eur-lex.europa.eu/legal-content/EN/TXT/?uri=CELEX%3A32014L0095

[69] California Department of Justice. California Transparency in Supply Chains Act, S.B. 657, 2010 Cal. Stat. Ch. 556. 2010. Available from: https://oag.ca.gov/SB657

Chapter 3

Perspective Chapter: Functional Applications in Textile Reuse and Recycling – Enhancing Sustainability and Reducing Environmental Impact

Ramratan Guru, Anupam Kumar, Jyoti Rani, Raul Fangueiro, Satyanarayan Panigrahi and Sakthivel Santhanam

Abstract

Used clothes and waste fibre materials are recycled and recovered using a number of processes to create recycled textile. The majority of recyclable textiles are found in municipal waste, which comes from items such as worn-out or abandoned clothing, tyres, shoes, carpets, furniture, non-durable products such as sheets and towels. The global demand for textile products is continually increasing as a result of population expansion and economic growth. This discussion includes assessments of studies on textile recycling and reuse. For these reasons, in this review chapter will be forgotten recycled textile materials based on some of the most recent main research papers at world level. In this chapter, various issues related to the area of environmental studies, new and innovative recycling process for polymers have been discussed on several aspect-related textured below: sustainable route; challenges and safety guides; methods used in order textile industry alone as well as polymer use application techniques. This chapter provides insights for researchers who are new in the area to by pointing out some of the iconic papers on this subject and enable them to get our hands-on various sources necessary along with a vision how they may propagate awareness about environment protection leading towards institutional commitment worldwide.

Keywords: reuse, recycling, textile materials, sustainability, impact on the environment

1. Introduction

The demand for fiber production has been made of this substance and is still growing because a brief 113 million-ton global fabric was generated in the summary year 2021 or, even if items continue as ordinary business, by ripping into 140 million tons end with the finish decade. Fast fashion has led to a significant waste buildup (with an estimated 92 million tons of textiles being wasted globally each year), as the demand for clothes increases, textiles are tossed away sooner than they used to be a generation ago. Of all this waste, less than 1% is recycled, but the rest ends up in landfills or being incinerated (approximately 73%), lost during production and collection (around 14%), or downcycled into lower-value applications. This leads to a considerable waste of valuable resources and severe environmental problems [1]. Fabric production continues to recover from the last few years, and world fabric manufacturing will continue to rise in 2021. All over the world, clothing has gone through a lot of trends and is in style from now on, but very short-period. But this ultimately shortens the lifecycle of that clothing before it goes to a landfill. Adjustable attire, which means constantly changing in clothing, leads to the rise of the textile industry, but it has disadvantages for generating large-scale waste on textiles. Today, waste textiles cannot be landfilled on a cost-effective basis because unlimited landfill space needs these days do not exist, and as large amounts deposited daily are exposed to rapid biodegradability of textile wastes in fanned atmospheres along with environmental hazards and costs related to their disposal in the form of garbage into landfills. So, using less textiles or recycling textile waste will help nature as well due to the decrease in non-renewable fossils that are used for making fibers. These reviews present an insight into different textile waste recycling technologies for fiber recovery from wastes into diverse products [2]. This study found that if textiles destined for reuse are collected separately from those for recycling, the environmental difference is dramatic and the impacts from reuse are lower than from recycling. It was observed that the manufacturing of textiles with recycled fibers is usually associated with lower environmental impacts compared to virgin fibers, except in several impact categories in the cases of cotton and polyester fibers [3]. This study also underlines the need for understanding the utilization of textile waste and by-products as precursors for functional materials and the availability of the potential of using textile waste in depollution applications as under-recognized. That has triggered the need to review the economic viability and the environmental impacts of moving from laboratory scale to pilot and industrial scale plants for commercial application [4]. The study focuses on exploring the potential of second-hand business by applying the mathematical model regarding environmental advantages of clothing reuse and taking Humana's case of Portugal. From 2016 to 2020, Humana in Portugal was able to sell 1426 tons of garments, which in return created an overall saving of 18,574,473 kg CO2eq emissions, 80,342,082 m^3 water consumption, and 121,941 MWh energy consumption. The study points out that clothing reuse has beneficial impacts on the environment and scourges the second-hand clothing situation in Portugal, particularly present that policy promotion might be helpful [5].

The paper focuses on the environmental concern of the textile and apparel industry through identifying its contributions to carbon emissions and effluent discharge as well as the emerging problem of textile wastes. The research is concerned with the fabrication of bricks out of textile waste, especially in construction and architectural design, through specimen testing with various binding agents. Finally, it analyzes and discusses the mechanical and thermal characteristics of these bricks used from reclaimed recycled textile waste in relation to the traditional clay bricks and introduces the concept of green interior design and construction [6]. The study notes the need to embrace systematic evaluations of environmental impacts linked to such new approaches in bio-recycling of textiles for their sustainability. By analyzing the Life cycle assessment (LCA) of the textile bio-recycling method, it was noted that the pre-treatment step was the largest fraction, trailing melt spinning and enzymatic hydrolysis [7]. This paper is thus concerned with the ineffective and expensive process of recycling waste fabrics, which has significantly contributed to environmental degradation and misuse of resources. The study presents a novel eco-friendly approach to recycling waste fabrics to fabricate versatile and energy-saving photothermal protective materials using MCFs coated with polydopamine and dodecylamine despite the poor photothermal performance of fabrics. The modified carbonized fibers (MCFs) derived from waste fabrics exhibited a light absorptivity of more than 95% in the visible region, heating up to 95 [8]. Specifically, the paper is devoted to the analysis of the consequences of apparel landfilling and recycling, noting that, worldwide, a considerable number of discarded clothes find their way to landfills whereas they can be recycled and used again. It includes data from research carried out in Australia, identifies comparative environmental impacts of landfilling natural apparel and synthetic apparel, and highlights the scope for environmental gains through recycling worn-out apparels [9]. This study zooms into the life cycle assessment of yarns produced from recycled cotton relative to virgin cotton in apparel production. It is, however, revealed that recycled forms of cotton yarns have relatively lower positions on life cycle environmental impacts than virgin products for almost all categories such as fossil fuel, abiotic, water, and land depletion, except for 62 and 95%. Of course, in terms of climate change and water depletion due to land occupation and due to irrigation water cotton, respectively [10]. This study describes an idea of a new article of the stitched sacrificial sewing thread that can be easily removed from this article of the clothing that will help it through the process of recycling the textile material [11]. The method adopted in the research paper is to recycle cellulose-containing textile material in the production of regenerated cellulose-molded bodies. The process includes, among others, the comminution of textile material, separating non-fiber components from fiber components, mechanical separation of non-cellulose fiber from cellulose fiber, chemical separation of additional non-cellulose fiber from cellulose fiber, and lastly, the formation of molded bodies from the cellulose fiber [12].

In this review chapter, we explored recycled textile materials by reviewing some of the most recent leading research papers globally regarding the topic. The chapter targeted a topic such as environmental studies, innovative applications of recycling, circular economy and a sustainable approach, challenges and precautions, textile recycling methods, as well as textile sectors' uses of recycled materials. This paper aspires to help new researchers highlight the studies in the field and also acquire various materials for raising environmental knowledge and advocating for sustainability around the globe.

2. Environmental benefits of textile reuse and recycling

2.1 Life cycle assessment (LCA) findings

It is now well documented that reuse and recycling of textiles is advantageous to the environment. Reuse is often found to create fewer effects on the environment than recycling. Generally, fabrics produced from recycled fibers entail less environmental

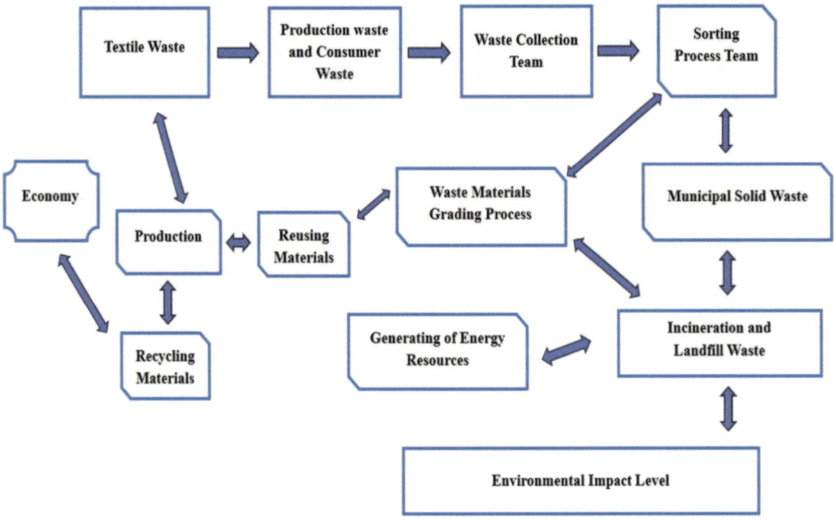

Figure 1.
Textile reuse and recycling process.

impacts than those from virgin fibers, other than where the fiber types are cotton/polyester [3].

2.2 Second-hand markets

The expanded use of second-hand clothing as a product type can be seen through the Portuguese case of Humana, to show how reuse is beneficial toward the environment. During 2016–2020, the operation of Humana has helped to prevent significant amounts of CO_2 emission, water usage, and energy consumption, which indicates that prolongation of garment life cycle is not only an operational and economic success but socially and ecologically advantageous as well (**Figure 1**) [5].

3. Innovative recycling applications

3.1 Functional materials from textile waste

Importantly, current textile waste is recyclable and can be used as functional materials for cleaning up the environment. For instance, waste wool is used to produce wool powder, which increases the efficiency of adsorbing the stocks of bad smells such as ammonia and formaldehyde gases [13].

3.2 Construction materials

Construction industry waste, such as textile waste, can be used in the production of construction bricks that should prove to be environmentally friendly. These bricks are produced from post-industrial waste textile materials and possess acceptable mechanical performance and thermal conductivity compared to standard materials, making them a feasible solution to decrease the ecological impact of construction [6].

3.3 Water decontamination

Some textile wastes can be recycled as adsorbents for water purification. These fibers, when modified, can function like filter papers in removing pollutants such as heavy metals and dyes from the water, which is part of management of the environment and conservation of resources [14].

4. Circular economy and sustainable practices

4.1 Bio-recycling and material recovery

A novel bio-recycling process opens up opportunities for recycling useful materials such as PET fibers and glucose from textiles, and the use of recycled materials instead of virgin products reduces the environmental impacts. But it is vital to apply optimization for these processes to increase energy effectiveness and material characteristics [7].

4.2 Construction applications

The textile waste can then be used in construction, for instance, in the making of bricks. These bricks, which are made out of textile fiber waste, have similar mechanical and thermal performances to traditional material, ensuring a circular textile system [6].

5. Challenges and considerations

5.1 Energy and resource efficiency

This means that the environmental advantages of recycling will depend on the efficiency of all the processes involved. For example, while recycling of bio-cotton-polyester blends is possible, certain processes, such as the pre-treatment, are very exhaustive lest the gains be eroded. It is therefore vital to find ways of enhancing energy efficiency in these processes in order to realize the most gains for the environment [7].

5.2 Economic and policy support

As evident from the study, there is a viable market opportunity for recycling of textile waste into value-added products; however, this can only happen after biting considerable economic analysis alongside innovative policies to support the sustainability of the business model. A greater focus invested in the collection and recycling of these materials has been shown to have profound benefits in minimizing the effects on the overall environment, but these should be backed up by the right policies and market signals [9]. Textile reuse and recycling make positive impacts in the environmental sector; still, there are hurdles that impinge on the potential of the processes, such as efficiency of processes, market-related factors, and policy support. The further implementations of advanced recycling applications and the supporting of second-hand markets are critical processes to move toward a sustainable textile industry.

5.3 Barriers to recycling

These challenges include complexity in fiber blends, contamination, and quality deterioration of fibers during recycling of the textiles. These issues inform the need to come up with advanced recycling technologies and methods in the near future [15, 16].

5.4 Value-added products

As it will be described below, an increasing focus on processing recyclable textiles into added value goods may add value to the entire process, thereby making recycling economically feasible. By adding these products to the market, the waste problem is solved while developing new market opportunities and taking care of the environment [17].

6. Methods of textile recycling

6.1 Mechanical and chemical recycling

On the basis of the processing technique used, the textile recycling can be classified into two categories, namely, mechanical and chemical recycling. In mechanical recycling, there are activities such as shredding and respinning of the fibers and the formation of new yarns and fabrics. The other type of recycling is known as chemical recycling, which entails dissolving the fibers into its chemical base for the production of new materials. Pyrolysis, enzymatic hydrolysis, and glycolysis methods are examined to demonstrate the possibility of recovering textiles, while pyrolysis and enzymatic hydrolysis have the issues of high price and technical difficulties (**Figures 2** and **3**) [18].

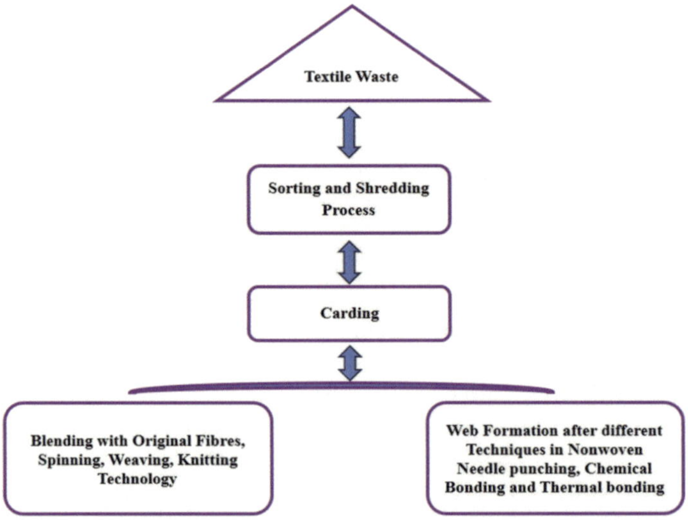

Figure 2.
Mechanical recycling process for textile waste.

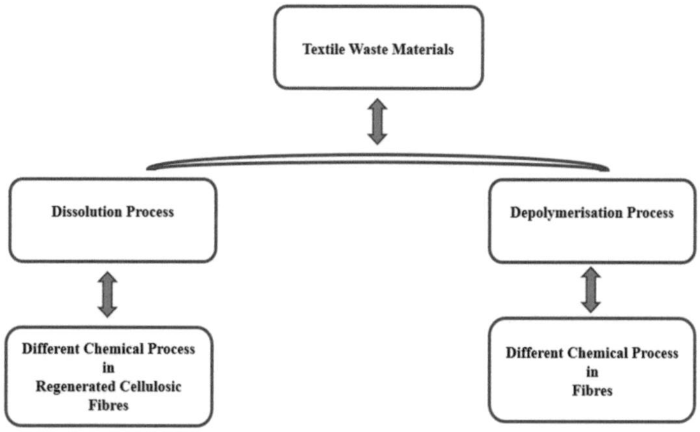

Figure 3.
Chemical recycling process for fibers.

6.2 Innovative techniques

New developments are the entrants in the sewing thread market that are sacrificial, that is, the threads useful in the sewing of garments in such a way that they make the recycling process easier. Furthermore, the IoT is also being considered for use in the sorting and identification of textile waste for further improvement of recycling [18].

6.3 Bio-chemical recycling

End-of-life textiles can be depolymerized using enzymes in a bio-chemical recycling process, breaking them down into their monomeric building blocks to serve as substrates for the production of value-added products like sugars (glucose) and bioethanol. The biological processes have the advantage of being low in energy demand, benign solvents and chemicals, and renewal instead of fossil carbon-based. Along with this, the process is extremely complex, which makes those perfect for separating combined textile waste [19, 20].

7. Economic and environmental impacts

7.1 Cost analysis

Economic viability of textile recycling depends on the cost of new as compared to recycled fiber. The cost of the recycled fibers and the cost of processing is normally what defines the feasibility of the recycling program. Recycling also has some disadvantage in that it requires more capital investment during the initial stages than the conventional method of disposal. However, if the long-term view is taken, this method proves cheaper in the long run since it involves lesser use of fill land and resources [21].

7.2 Carbon footprint reduction

Recycling of textiles has been proven to elicit a direct positive impact of decreasing the carbon footprint of textiles. There is an existing potential for emission reductions through recycling process waste and end-of-life products in the present structure of the industry. As highlighted by the previous LCA, a successful implementation of the recycling strategies has the capability of bringing down the carbon footprint by a great amount (**Table 1**) [15].

Authors	Practical implications	Results	Conclusions
Sanjrani Manzoor Ahmed et al. [22]	The developing nations have been depicted as having little or no systematic management of textile solid waste; hence, they call for proper infrastructure and policies that will address the disposal of the endangered textile waste.	It is evident that most developing countries have no systematic way of disposing of textile solid waste, making this a big challenge in disposing of the waste.	The bulk of these textiles end up as solid waste due to poor waste management in developed countries; developing countries still lack systematic waste management for textile solid waste.
Fatma Şener Fidan et al. [23]	The paper also concludes that there is a need to use recycled raw materials in the manufacturing of the denim fabric, especially the recycled PET fiber, in order to minimize the environmental impacts, including freshwater ecotoxicity, eutrophication, and acidification. This has a close connection to global textile industry practices to adopt eco-friendly production techniques so as to reduce its contribution to global warming.	The analysis of the life cycle implication also defined that recycled PET fiber has less negative effect having regard to ecotoxicity using freshwater, eutrophication, and acidification than traditional PET fiber and also than cotton fiber.	It emerged from life cycle assessment that recycled PET fiber yields considerably lower freshwater ecotoxicity than both conventional PET fiber and cotton fiber in addition to giving less eutrophication and acidification.
Torun Hammar et al. [24]	This is true when selecting a chemically recycled cellulose carbamate fiber for garment production since it had a low total climate impact, a low total water scarcity impact, and a low total land use impact compared to primary conventional cotton and organic cotton.	All considered environmental impacts were reduced by applying recycled cellulose carbamate fiber in place of primary conventional cotton, amounting to a positive environmental impact effect of the chemically recycled fiber.	Comparing the environmental impacts of chemically recycled cellulose carbamate fiber to primary fibers demonstrated lower overall boundaries for the evaluated garment; however, there are trade-offs in different categories of environmental effects between different fiber types.
Kaushik Roy Choudhury et al. [25]	The paper emphasizes the issue of disentanglement of cotton, polyester, and elastane fibers in blended textiles for better opportunities for their recycling while reducing environmental impacts of the textile industry.	Some of the problems with blended textiles being mechanically recycled are due to the inability to separate the blend from the fibers.	Due to this, the paper concludes that for sustainable strategies in textile recycling, there is a need to separate constituents that are important in blended textiles, such as cotton, polyester, and elastane.

Authors	Practical implications	Results	Conclusions
Anabela Gonçalves et al. [26]	The paper also focuses on the necessity of using life cycle assessment and such tools as open LCA along with the coinvents database for the proper evaluation of the environmental impact in the textile sector.	A Portugal-based white T-shirt made from 50% post-industrial waste and recycled cotton and 50% organic cotton from Turkey was examined in the study concerning its environmental impact. Key among the findings was that the use of recycled cotton from Portugal emerged as having some environmental advantage over the two categories of cotton.	The study concludes by supporting the environmentally positive impact of introducing postconsumer recycled cotton, which is produced in Portugal and is produced by repurposing post-industrial waste for textile manufacturing.
Erha Andini et al. [1]	Therefore, the chemical recycling method described in the paper addresses the problem of a low recycling rate of postconsumer textile waste by converting mixed fibers into their monomers for the subsequent production of new textiles without the use of virgin materials.	The utilization of the postconsumer mixed fabric waste comprising of polyester and spandex is done using the microwave-assisted glycolysis catalyzed by ZnO followed by solvent dissolution, where the material turns into glycolide in 15 minutes.	According to the findings presented in the paper, the chemical recycling of the mixed textile waste using microwave-assisted glycolysis over ZnO catalyst and solvent dissolution is a commercially viable process for the rapid depolymerization of the PET and spandex to their monomers in about 15 minutes.

Table 1.
Presented are the aggregated findings from studies on the recycling and reused textile and circular economy.

8. Recycling material in textile sectors

The recycling of textile materials is well established, both in open and closed loops. Despite being an ideal manufacturing hub on every parameter, be it price, volume, or the availability of raw materials, there are challenges faced in building economically operational units. Fashion in particular has demonstrated the ability to perform open-loop recycling. Furthermore, open-loop recycling is a system through which textile waste can be recycled into other products/applications, while closed-loop recycling only turns recyclable products into the same type of their original form. The majority of textile waste with its mix over fiber type and color leads to shoddy fibers, which have limited visual appeal and are unsuitable for being respun into quality yarn (closed-loop recycling). As a result, recycled textiles are only really useful as an input for low-margin products such as carpet underlay or insulation material [27].

8.1 Recycling consequences of textile blends

In order to attain the necessary properties needed for clothing usage, natural fibers are often blended with regenerated or synthetic fibers. Example of further fiber blending with natural or regenerated fibers (cotton blend, polyester blend, lycra blend, viscose blend, linen blend, and lyocell blend) percentage-wise. In the textile production sector, blending is an accompaniment to this; it occurs at various stages of manufacturing and makes recycling complex as fiber blends in waste textiles need separating and sorting into their constituent fibers. As a result, the vast majority of textile waste ends up in landfills with severe ramifications for all essential resources. Usually, they are made of multiple fibers and single-fiber polymer materials, making it difficult to recycle.

8.2 The impact of fabric chemical treatments on recycling

The textile production processes use a vast range of chemicals and water to help the methods that are used in achieving the required properties for these textiles. Chemicals are added to fabric by the dyeing, printing, and finishing processes that remain on textiles during their use. Dyeing and Printing: Processes of adding color and prints to the fabric with dyes or pigments (artisans factory). Chemical finish applications can also be used for such technical properties as antimicrobial, anti-wrinkle, and fabric softeners and flame retardants, depending on the materials and textile fabrics. Fabric chemical treatments exercise dyes or purposeful agents; however, a number of the chemicals continue to be in textile products at end-of-life even after multiple launderings [28].

Textile recovery and recycling, chemical bond dye fiber. Nevertheless, methods are available in brief for the removal of different dye classes. Chemicals degrade chemical recycling, so we need pre-treatment steps. Various dyes and other pollutants that usually sublime or coagulate into insoluble particles can cause complications in the respective respining techniques [29].

8.3 Dashing for trash: Picking up waste and segregating

The main function of waste recycling is a correct separation and identification by the composition of individual objects. Textile material identification: Textiles are usually composed of various components, such as zippers and buttons or other decorations, making it quite difficult to identify textile materials in the process of textile recycling since they have a very complex structure. Textiles are currently manually sorted and identified one by one, which not only makes handling large volumes of material a limitation but also does not lend itself to full automation. Identification of textile materials using microscopy has been reported. Greater challenges are often encountered in discriminating at the level of species within a genus from morphological data, e.g., identification of an assemblage known to contain cotton from one that corresponds instead with viscose [30].

9. Basically, recycled fiber and yarn future scenario

9.1 Fiber to yarn process

Over three-quarters (65–76%) of global textile recycling income came from cotton materials in 2021. This segment is expected to grow at a CAGR of 2.5% during the forecast period. One of the main materials in clothing and other textiles is cotton; thus, it gets discarded a lot. A lot of the recycled cotton is one-third virgin or non-renewable treated fabric. Wool clothing is inherently strong and can last longer in the textile industry. Unspun wool or recycled wool is really easy to recycle, and so, in Prato, Italy, for example, lots of recycling goes on. A study by the International Wool Textile

Organization (IWTO) reveals that remnant wool can make a difference to radicalize the market if correctly used and cared for. There is a growing need in the clothing industry, so polyester recycling gains momentum. Corporate said they start to make more clothes of really recycled polyester instead of regular ones (**Figures 4** and **5**) [31, 32].

9.2 Textile waste to fabric developed

Apparel waste held a 28.75% share in the revenue generated from textile recycling across the globe and is estimated to advance at 3.1% CAGR during the forecasting horizon of the comparison year (2020–2026). Visibly, the trash is clothing scraps, defective garments, and shoes the owner has thrown away, along with leftover fabric from production. During the past two decades, new clothes have lost nearly a year off their average lives. Just 5 years ago, Europeans bought 40% more clothing items than they do today, and retail prices of apparel fell by around one-fifth. Cushions, carpets, and rugs, mattresses, coverings with fabrics cousin curtains sofas. This unfortunately led to more waste as disposable incomes globally increased, leading people worldwide to purchase them (**Figure 6**) [31].

(a)

(b)

Figure 4.
(a) Recycling fiber with original fiber mixing. (b) After refinement with mixing for recycling fiber.

Figure 5.
Recycling fiber to yarn package.

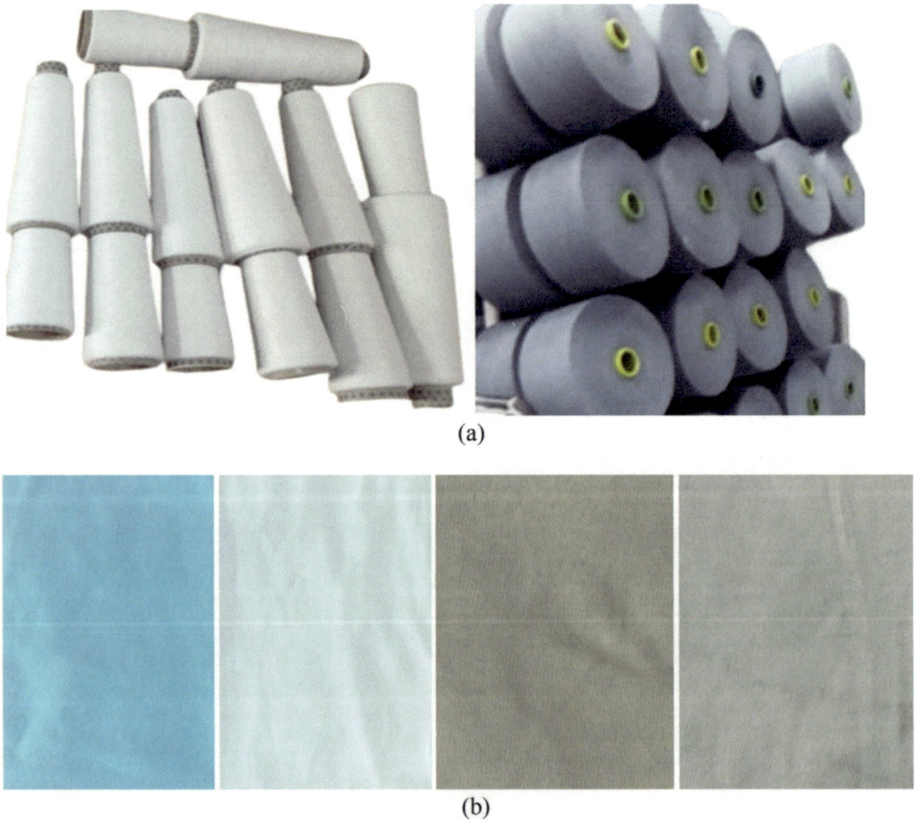

Figure 6.
(a) Recycling yarn packages. (b) Recycling fabrics.

Textile recycling is better for the environment as it reduces energy and water waste, saves on virgin fibers, which decreases spinning yarn production emissions, uses fewer chemical dyes that pollute land soil and river waters, etc. More recently, it has begun to demonstrate how big a thing sustainability can be in the clothing industry. For these reasons, recycled textiles are applied across a variety of sectors, including apparel, home furnishings, retail automotive, mining, building, and construction.

10. Conclusions

The paper on Functional Application for Textile Reuse and Recycling to Impact of Environment reveals that reuse and recycling of textiles have a vital role in minimizing the impact of the fashion and textile industries on the environment. Reducing the amount of freshly produced cloth and trash that ends up in landfills by practicing the reuse of cloths as well as by recycling the old and discarded clothes lowers landfill waste and resource utilization and reduces greenhouse gas emissions as well. The study claims that further development of recycling technologies and introducing a fast and efficient system of collecting and sorting what is to be recycled will provide the highest opportunities. Besides, it also suggests the necessity of higher consumer activation and involvement, as active consumer involvement is one of the important success factors of such initiatives. Thus, the study brings attention to the fact that recycling has to become a or, in fact, remain a pivotal focus in textiles as an attempt to reduce environmental effects and attain sustainable development objectives in the long run.

Author details

Ramratan Guru[1*], Anupam Kumar[2], Jyoti Rani[3], Raul Fangueiro[4], Satyanarayan Panigrahi[5] and Sakthivel Santhanam[6]

1 School of Design, Mody University of Science and Technology, Lakshmangarh, Rajasthan, India

2 Giani Zail Singh Campus College of Engineering and Technology, MRSPTU, Bathinda, Punjab, India

3 School of Fashion, Lovely Professional University, Phagawara, Punjab, India

4 School of Engineering, University of Minho, Guimaraes, Portugal

5 Research Chair Innovative Manufacturing, School of Mining Energy and Manufacturing, Saskatchewan Polytechnic, Saskatoon Campus, Canada

6 School of Design, Presidency University, Bangalore, Karnataka, India

*Address all correspondence to: ramratan333@gmail.com

References

[1] Andini E, Bhalode P, Gantert E, Sadula S, Vlachos DG. Chemical recycling of mixed textile waste. Science Advances (American Association for the Advancement of Science). 2024;**10**(27):1-12

[2] Rivalani BB, Gbadeyan OJ, Sithole B, Chunilall V. Recent advances in recycling technologies for waste textile fabrics: A review. Textile Research Journal. 2024;**94**(3-4):508-529

[3] Abagnato S, Rigamonti L, Grosso M. Life cycle assessment applications to reuse, recycling and circular practices for textiles: A review. Waste Management. 2024;**182**:74-90

[4] Tummino ML, Varesano A, Copani G, Vineis C. A glance at novel materials, from the textile world to environmental remediation. Journal of Polymers and the Environment. 2023;**31**(7):2826-2854

[5] das Virgens N, Silva SC e, Laranjeira EF. Applications of the circular economy to the second-hand textile and clothing market: The case of Humana in Portugal. International Journal of Fashion Design, Technology and Education. 2022;**16**:214-223

[6] Jamshaid H, Shah A, Muhammad Shoaib RK, Mishra. Recycled-textile-waste-based sustainable bricks: A mechanical, thermal, and qualitative life cycle overview. Sustainability. 2024;**16**(10):1-15

[7] Subramanian K, Chopra SS, Cakin E, Li X, Lin CSK. Environmental life cycle assessment of textile bio-recycling valorizing cotton-polyester textile waste to pet fiber and glucose syrup. Resources Conservation and Recycling. 2020;**161**:104989

[8] Li Y, Li H, Jun W, Jia X, Zhang Z, Yang J, et al. Facile recycling of waste fabrics for preparing multifunctional photothermal protective materials. ACS Sustainable Chemistry and Engineering. 2023;**11**(28):10566-10577

[9] Moazzem S, Wang L, Daver F, Crossin E. Environmental impact of discarded apparel landfilling and recycling. Resources Conservation and Recycling. 2021;**166**:105338

[10] Liu Y, Huang H, Zhu L, Zhang C, Ren F, Liu Z. Could the recycled yarns substitute for the virgin cotton yarns: A comparative LCA. International Journal of Life Cycle. 2020;**25**:2050-2062

[11] Serbiak PJ. Patent: Method for Recycling Textile Material. Taiwan: WIPO (PCT); 2019

[12] Christoph K-N, Richard H, Christian W. Patent: Process for Recycling Cellulose Containing Textile Material. Taiwan: Patent; 2019

[13] Tang W, Tang BWY, Bai W, Pakdel E, Wang J, Wang X. Porous, colorful and gas-adsorption powder from wool waste for textile functionalization. Journal of Cleaner Production. 2022;**366**:132805-132805

[14] Bediako JK, Apalangya VA, Hodgson IOA, Anugwom I, Repo E. Adsorbents for water decontamination: A recycling alternative for fiber precursors and textile fiber wastes. Science of the Total Environment. 2024;**1**(919):171000

[15] Muthu SS, Li Y, Hu JY, Ze L. Carbon footprint reduction in the textile process chain: Recycling of textile materials. Fibers and Polymers (The Korean Fiber Society). 2012;**13**(8):1065-1070

[16] Wang Y, Zhang Y, Polk MB, Kumar S, Muzzy JD. Recycling of Carpet and Textile Fibers. John Wiley and Sons, Inc.; 2005. pp. 697-725

[17] Hawley JM. Textile Recycling. Book: Handbook of Recycling. Elsevier Inc.; 2014. pp. 211-217

[18] Damayanti D, Wulandari LA, Bagaskoro A, Rianjanu A, Ho-Shing W. Possibility routes for textile recycling technology. Polymers (Multidisciplinary Digital Publishing Institute). 2021;**13**(21):3834

[19] Yunzi H, Chenyu D, Leu S-Y, Jing H, Li X, Lin CSK. Valorisation of textile waste by fungal solid-state fermentation: An example of circular waste-based biorefinery. Resources, Conservation and Recycling. 2018;**129**:27-35

[20] Piribauer B, Bartl A. Textile recycling processes, state of the art and current developments: A mini review. Waste Management and Research. 2019;**37**(2):112-119

[21] Wang G. Evaluation System's Research on Textile Fiber Material Ecological Recycling. Advanced Materials Research. Vol. 1055. Switzerland: Trans Tech Publications; 2014. pp. 92-95

[22] Ahmed SM, Xue G, Mirza SNA. A review on textile solid waste management: Disposal and recycling. Waste Management and Research: The Journal for a Sustainable Circular Economy. 2024;**42**(7):1-18. DOI: 10.1177/0734242X241257093

[23] Fidan FŞ, Aydoğan EK, Uzal N. Sustainability assessment of denim fabric made of PET fiber and recycled fiber from postconsumer PET bottles using LCA and LCC approach with the EDAS method. Integrated Environmental Assessment and Management (Wiley). 2024;**20**(6):2347-2365

[24] Hammar T, Peñaloza D, Hanning A-C. Life cycle assessment of a circular textile value chain: The case of a garment made from chemically recycled cotton. International Journal of Life Cycle Assessment (Springer Science, Business Media). 2024;**29**(10):1880-1898. DOI: 10.1007/s11367-024-02346-2

[25] Choudhury KR, Tsianou M, Alexandridis P. Recycling of blended fabrics for a circular economy of textiles: Separation of cotton, polyester, and elastane fibers. Sustainability. 2024;**16**(14):6206-6206

[26] Gonçalves A, Baêta M, Nagy M, Silva C. Looking for environmental scoring: A case study of a Portuguese cotton white T-shirt made with recycled Fiber. Sustainability. 2024;**16**(14):5896-5896

[27] Ribul M, Lanot A, Pisapia CT, Purnell P, McQueen-Mason SJ, Baurley S. Mechanical, chemical, biological: Moving towards closed-loop bio-based recycling in a circular economy of sustainable textiles. Journal of Cleaner Production. 2021;**326**(1):129325

[28] Schmidt C, Berghahn E, Ilha V, et al. Biodegradation potential of Citrobacter cultures for the removal of amaranth and Congo red azo dyes. International journal of Environmental Science and Technology. 2019;**16**:6863-6872

[29] Määttänen M, Asikainen S, Kamppuri T, Ilen E, Niinimäki K, Tanttu M, et al. Colour management in circular economy: Decolourization of cotton waste. Research Journal of Textile and Apparel. 2019;**23**(2):134-152

[30] Nayak R, Houshyar S, Khandual A, Padhye R, Fergusson S. Identification of natural textile fibres. In: Kozłowski RM, Mackiewicz-Talarczyk M, editors. Handbook of Natural Fibres. Duxford, United Kingdom: Elsevier; 2020. pp. 503-534

[31] Guru R, Rani J, Yadav P, Kumar S, Saini U. Basic application for fibre and fabric recycling to sustainability products. Shodh Drishti (An International Peer Reviewed Referred Research Journal). 2023;**14**(8.1):183-190

[32] Guru R, Kumar A, Grewal D. The recycling and reuses plastic bottles to fibre process and applications. Man Made Textile in India. 2022;**L**(6):162-166

Chapter 4

AI-Based Width Measurement System

Muhammad Mirza Nouman Baig

Abstract

This chapter presents the development and evaluation of an AI-based fabric width measurement system designed to enhance the accuracy and efficiency of fabric production processes. Utilizing the YOLOv5s algorithm, the system captures real-time fabric width measurements, displaying them on both an human machine interface (HMI) and a web interface. The experimental setup, which includes data labeling, epoch size, and batch size, is thoroughly detailed, demonstrating the model's ability to achieve high precision and reliability. Performance metrics such as train/box_loss, train/obj_loss, train/cls_loss, and metrics/mAP_0.5:0.95 are analyzed to assess the system's effectiveness. The final output confirms the model's capability to provide accurate fabric width measurements with an impressive accuracy rate of approximately 90%. Additionally, the system's real-time operational potential and future integration with machine PLCs for enhanced control over fabric shrinkage are discussed. This AI-based solution offers significant improvements over traditional manual measurement methods, promising advancements in quality control and operational efficiency in the textile industry.

Keywords: AI-based fabric measurement, YOLOv5s algorithm, machine learning, real-time measurement, textile industry, performance metrics, quality control

1. Introduction

The global economy depends on the textile industry, which makes a considerable contribution to employment and trade on a global scale. It includes the manufacturing of numerous textiles and textile goods using different techniques, such as weaving, dying, printing, and finishing. To stay competitive in the market, the industry works to continually increase its operational effectiveness, product quality, and cost-effectiveness. The deployment of automation and technology has significantly advanced in the textile sector in recent years. Textile producers have embraced creative solutions to fulfill the rising customer demands and address operational issues. In the textile sector, AI has become a potent instrument for operational modernization. The manufacture of textiles has benefited from the use of AI techniques like machine learning and image analysis. Quality control, defect identification, production optimization, and supply chain management are just a few of the operations that can be automated thanks to these methods. Textile producers may increase production, raise product quality, lower prices, and better adapt to market demands by utilizing AI.

1.1 Current AI-based solutions

1.1.1 Defect detection system

The use of AI-powered defect detection systems is a noteworthy AI initiative in the textile industry. These systems employ deep learning methods and computer vision algorithms to automatically recognize and categorize fabric flaws like stains, holes, and unusual patterns. Manufacturers can swiftly discover and address quality concerns, saving waste and raising the caliber of their finished products, by automating the defect detection process [1].

1.1.2 Predictive maintenance

Predictive maintenance for textile machinery is the topic of another AI-based research. Manufacturers can track the functioning of machinery in real time and anticipate probable faults or maintenance requirements by utilizing machine learning algorithms and sensor data. This preventative strategy enables prompt maintenance interventions, reducing downtime and increasing production effectiveness [2].

1.1.3 Inventory management solutions

The textile industry has also seen a rise in the use of AI-driven inventory management solutions. In order to optimize inventory levels and guarantee timely material availability, these systems use AI algorithms to analyze past sales data, demand forecasts, and supply chain variables. Manufacturers can reduce stockouts, carry costs, and customer satisfaction by precisely forecasting demand and optimizing inventory [3].

1.1.4 Optimizing production process

Additionally, scheduling and planning textile production with AI has proved beneficial in streamlining and improving production methods. AI algorithms are able to create effective production plans that minimize downtime, maximize resource utilization, and shorten lead times by taking into account a variety of factors, including order volumes, machine capacities, and production restrictions [4].

1.1.5 Color and design identification

The industry has also benefited greatly from AI-based technologies for textile pattern and color identification. These systems use machine learning algorithms to analyze and classify the patterns and colors in fabrics, giving designers and producers the ability to swiftly spot trends, produce complicated patterns more quickly, and develop bespoke designs [5].

2. Objectives and scope of the project

The design and implementation of an AI system specifically suited for fabric width measuring on sanforizing machines is the main goal of this research. The project wants to accomplish the following goals:

- Create a strong AI model based on image analysis methods that can precisely measure cloth width.
- To do proper real-time model testing on fabric.
- Design a user-friendly interface for tracking and visualizing cloth width measurements in real time.
- Compare the AI system's efficiency and precision against conventional measurement techniques.

The project's scope includes the full design and implementation of the AI system for calculating fabric width. This involves a number of steps, such as the collection of the dataset, labeling of images, training of the AI model, integration of the model with the webpage and database, and performance testing of the system. It is crucial to remember that the project only addresses the measurement of fabric width; it does not cover any other facets of sanforizing or textile manufacturing processes.

The textile industry is a cornerstone of global manufacturing, where precision and efficiency in fabric production are crucial. Traditional methods for measuring fabric width are often manual, time-consuming, and prone to human error. To address these limitations, this chapter presents an advanced AI-based fabric width measurement system, utilizing machine learning algorithms to enhance accuracy and streamline operations. By integrating the YOLOv5s algorithm, this system offers real-time fabric width measurement, significantly improving process control and quality assurance in textile manufacturing.

3. Methodology

Basically, methodology identifies the applied methods to find a proper and complete solution to an outlined problem. It allows us to get a detailed plan that helps researchers on track, making the process smooth, effective, and manageable. Methodology will enable us to get structured steps to solve a problem which will ensure that the system is working efficiently.

3.1 Dataset creation

This is the very first step in developing a customizable machine learning model. Initially, we created a dataset by capturing 500 images of fabric. Then all the jpeg files are renamed according to the numbering of images. Creation of the dataset includes the two main tasks mentioned below:

3.1.1 Data labeling

Data labeling is the process of tagging data with relevant information to form a labeled dataset. In this AI-based width measurement system data labeling refers to the process of assigning the desired detection area in the images. All the images are uploaded to the makesense.ai tool for labeling. By using the rectangle tool, the desired area in each image is identified manually. All the images are labeled one by one to complete the overall data labeling of the custom dataset. The process of data labeling is shown below (**Figure 1**):

3.1.2 Data annotation

The process of adding metadata or labels to raw data to offer more context, information, or meaning is known as data annotation. The machine learning algorithm understands this data annotation which includes labeling, tagging, or any parameter or properties in the dataset.

When we done the data labeling step in makesense.ai, it gives the text file of an individual image which holds some relevant information regarding data labeling. So now we have downloaded all the text files of images from sensemake.ai.

3.2 Machine learning model

3.2.1 Model selection

YOLOv5, a prominent object detection algorithm in computer vision and machine learning, is employed in the AI-based fabric width measurement system. When the project commenced, YOLOv5 was highly demanded and user-friendly. Known for its speed, efficiency, and high detection accuracy, YOLOv5 met the core requirements of the project. Its customizable design allows for adaptation to specific application needs, enabling fine-tuning on the fabric width dataset to precisely recognize fabric edges and boundaries. Integration with the database and website *via* API is straightforward. Additionally, the algorithm's strong, active community offers extensive resources, tutorials, and pre-trained models, providing robust support for troubleshooting during deployment. Therefore, YOLOv5 is an optimal choice for addressing the industry's fabric width measurement challenges.

3.2.2 Model creation

As we have discussed above the machine learning model selected for this project is Yolov5, so the built-in Yolov5 model is downloaded from official GitHub and extracted in the PC with the folder named "Yolov5." In the directory Yolov5 we have created two folders named "train" and "val." In the train and val folder, two additional folders called "images" and "label" are created.

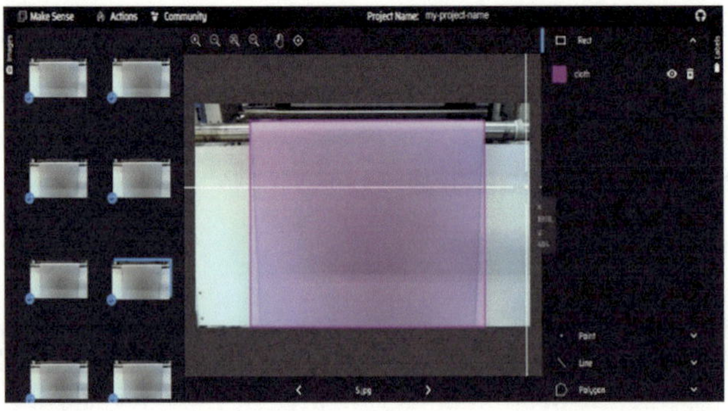

Figure 1.
Data labeling sample.

Now the dataset created is divided into three parts w.r.t folders discussed below:

3.2.2.1 Train

In this folder, we have placed 70% of the dataset in the form of jpeg and text files in the folder of images and labels respectively.

3.2.2.2 Val

In this folder, we have placed 20% of the dataset in the form of jpeg and text files in the folder of images and labels respectively.

Lastly the remaining 10% of the dataset is kept aside for future model testing phases.

Furthermore, there are multiple steps that need to be followed to completely deploy the machine learning model in real time. The steps are provided below:

3.2.2.3 Libraries and dependencies installation

There are some dependencies that need to be installed before the model training. In a folder named "data" there is a yaml file named "coco128.yaml," the purpose of this file is to load the path of our customized dataset and identify the classes of the dataset. This file is updated according to our dataset directory and class. As there is only one class in our dataset which is width.

Furthermore, we have downloaded the ylov5l6.pt. file from GitHub and extracted in the "Yolov5" folder. The purpose of this file is to store the learned weights and parameters of a Yolov5 model that has been trained on a specific dataset.

There is a virtual environment created on PC for machine learning model training. This command is used in cmd to install the virtual environment library:

Virtualenv myenv

After this, the environment is activated using this command:

Myenv\Scripts\activate

After that, the model-required libraries are installed using this command:

pip install –r requirements.txt

Till now the major libraries and dependencies are installed in PC. So, now the system is ready to train the machine learning model.

3.2.3 Model training

As it is, make sure that the environment is created successfully and all the libraries and dependencies are installed perfectly, so we can head toward model training. On cmd, we should be in "Yolov5" directory, and after that, the machine learning model is trained through a single command given below:

!python train.py --img 1280 --batch 16 --epochs 10 --data coco128.yaml --weights yolov5l6.pt –cache

Here the command is described precisely for understanding:

!python: The exclamation mark (!) at the beginning of the command indicates that it is being run within a Jupyter Notebook or a similar interactive environment.

train.py: Refers to the Python script being executed. In this case, it is a script named "train.py" that is likely responsible for training a YOLOv5 model.

--img 1280: Specifies the image size to be used during training. In this case, the value is set to 1280 pixels.

--batch 16: Sets the batch size for training. The value is set to 16, indicating that 16 images will be processed in each training iteration.

--epochs 10: Specifies the number of epochs or training iterations to be performed. The value is set to 10, indicating that the training process will iterate over the entire dataset 10 times.

--data coco128.yaml: Indicates the YAML file containing the configuration and dataset information. In this case, the file is named "coco128.yaml" and likely contains details about the dataset used for training.

--weights yolov5I6.pt: Specifies the weight file to be used for training. The value is set to "yolov5I6.pt," indicating that the pre-trained weights from that file will be loaded before training.

--cache: Enables caching of images for faster training. It allows the model to store preprocessed images in memory, reducing disk I/O during training.

Furthermore, in the detect.py file there is an addition mentioned below:

```
for *xyxy, conf, cls in reversed(det):
c1, c2 = (int(xyxy[0]), int(xyxy[1])), (int(xyxy[2]), int(xyxy[3]))
center_point = round((c1[0] + c2[0])/2), round((c1[1] + c2[1])/2)
circle = cv2.circle(im0, center_point,5,(0,255,0),2)
text_coord = cv2.putText(im0, str(center_point), center_point, cv2.
FONT_HERSHEY_PLAIN,2,(0,0,255))
print (center_point[0]/16.43)
```

This code is actually measuring the width of the fabric, as the model was detecting in the form of a bounding box so we took the parameters of the box, and on the basis of those numbers, we scaled the number to get it converted into the inches. The bounding box is in the xyxy array, which we converted into an integer, and after rounding the data, we divided this by 16.43 so that we have achieved the real-time width of fabric in inches.

3.2.4 Sample testing

After this command the machine learning model is trained and the weights of this model are created. So now the model is going to be run on PC using cmd with the command given below:

python detect.py –weights runs\train\exp\best.pt –conf 0.7 –img 1280 –source sample.mp4

Here the command is described precisely for understanding:

python: Refers to the Python interpreter that will execute the script.

detect.py: Specifies the Python script that contains the code for object detection.

- **weights runs\train\exp\best.pt**: Specifies the path to the weights file of the trained model. In this case, the file is located at runs\train\exp.\best.pt. The weights file contains the learned parameters of the trained model.

- **conf 0.7**: Sets the confidence threshold for object detection. The value is set to 0.7, indicating that only detections with a confidence score greater than or equal to 0.7 will be considered.

- **source sample.mp4**: Specifies the source of the input for object detection. In this case, the source is a video file named sample.mp4. The script will process the frames of the video to perform object detection.

3.2.5 Real-time testing

As the model had been tested on a sample video and detected the width perfectly. Now the model has been tested on real-time production of fabric. While the fabric is passing the inspection table, the model detects the fabric in real-time and it shows the width of the fabric on-screen live. The physical width of the fabric is accurately equaled to the width predicted by this model. The model has shown an accuracy of 80% in predicting the width of the fabric. The system has performed very efficiently and accurately in measuring the width of the fabric.

3.2.6 Model deployment

As the machine learning model is completely tested in real-time, now it can be deployed permanently on the machine for operators' ease. The hardware is installed perfectly on the machine. The model is running on a PC, and the user interface for operators is remaining which is the next part. Till now the AI-related work has been deployed on this system.

3.3 Website development

3.3.1 Frontend development

Frontend development involves creating the graphical user interface (GUI) using JavaScript, CSS, HTML, jQuery, and Bootstrap. The main files are **content.php**, which serves as a static homepage with introductions and images; **live.php**, which displays real-time fabric width measurements from the machine learning model and includes fields for monitoring and entering fabric lot numbers; **signup.php**, which allows authorized personnel to create accounts with fields for personal details; and **login.php**, which provides a form for users to access their accounts using email and password (**Figures 2–5**).

3.3.2 Backend development

Backend development, also known as server-side development, is responsible for the behind-the-scenes logic that powers the web application. In this project, the backend integrates the system's database with the front end, ensuring seamless data flow and real-time display of fabric measurements on the webpage. The width of the fabric shown on the webpage is working on this file. Basically, the data is coming through API in the webpage from the database. The **"fetchapi.php"** file fetches data to the webpage from the database.

3.4 Database development

The goal of database development is to design a framework that will make it possible to store and retrieve data quickly. Some of the basic advantages of the database in this system are to analyze data in a variety of ways, promote a disciplined approach to data management, turn disparate information into a valuable resource, and improve the quality and consistency of information of the machine (**Figure 6**).

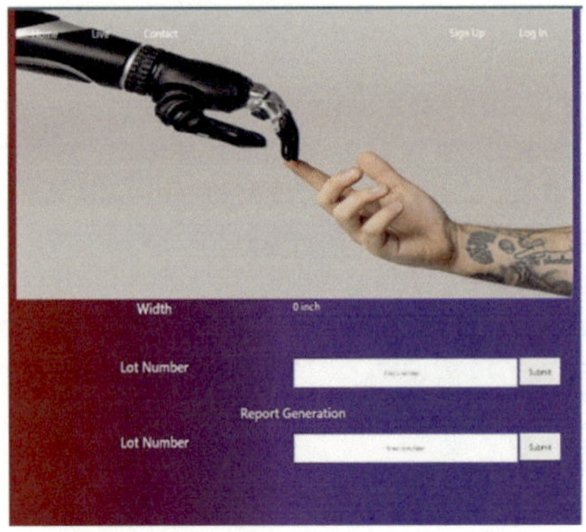

Figure 2.
Live page demonstration.

Figure 3.
Content page demonstration.

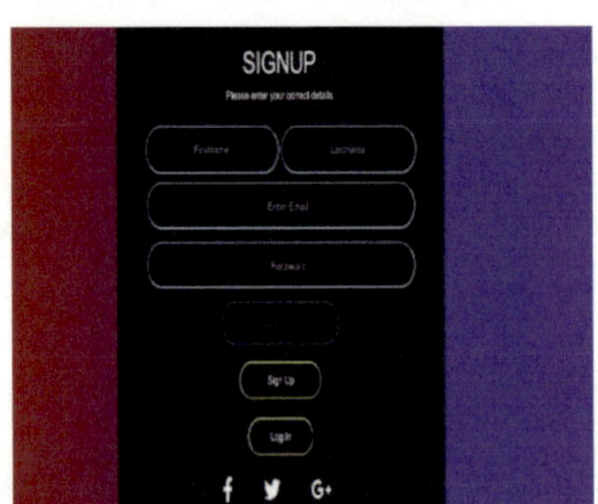

Figure 4.
Signup page demonstration.

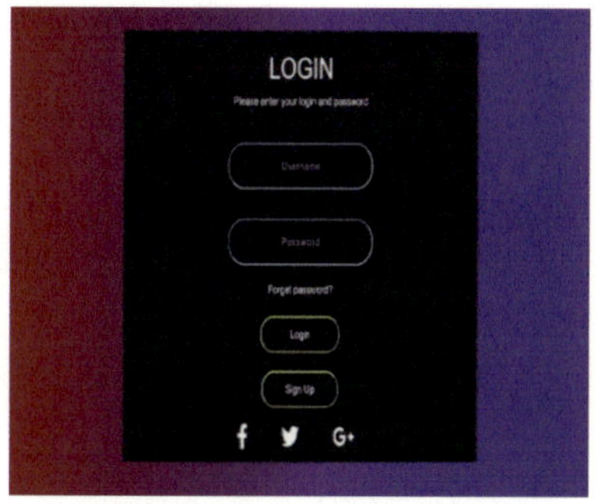

Figure 5.
Login page demonstration.

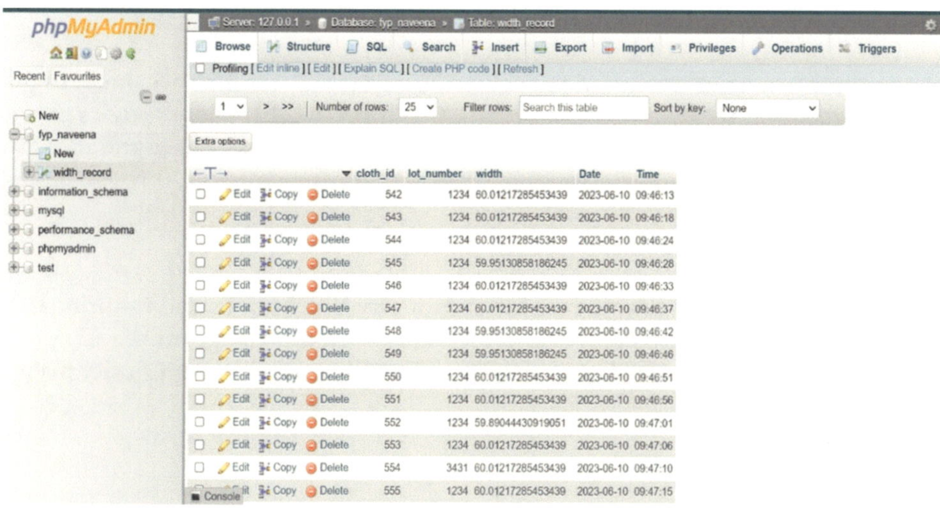

Figure 6.
Database demonstration.

MYSQL language is used for database creation. A database is created named "fyp_naveena"

3.4.1 Record table creation

As the operators demanded that they need the lot number, width, time, date, and cloth id. So, there are five columns made in the fyp_naveena database through XAMPP software. The table created is named "width_record."

3.4.2 API development

API stands for application programmable interface. API is generically used to describe connectivity interfaces to an application. We need to use API in our project to integrate model data with the database and website.

Initially, we created a folder named **"api."** In **"api"** folder we made two files named **"insertapi.php"** and **"fetch.php."** The purpose of insertapi file is to pick data from the machine learning model and save it to the database. The fetchapi file is used to send model data to the website from the database.

4. Experimental design study

4.1 Performance metrics

Performance metrics are used to calculate the effectiveness and accuracy of the AI-based fabric width measurement system. The importance of performance metrics in AI-based width measurement systems is described below briefly:

4.1.1 Quantitative evaluation

Quantitative metrics were used to evaluate the system's performance, including accuracy, precision, recall, and mean average precision (mAP). These metrics provide a comprehensive assessment of the model's ability to detect fabric edges and measure width accurately.

4.1.2 Comparison with benchmarks

The AI-based system was compared with industry benchmarks and traditional measurement methods. This comparison highlighted the advantages of the AI system in terms of accuracy, speed, and efficiency, demonstrating its superiority over conventional methods.

4.1.3 Quality control and process optimization

Performance metrics play a crucial role in quality control and process optimization. By analyzing these metrics, manufacturers can identify areas for improvement, optimize production processes, and enhance fabric quality.

4.2 Decision-making and system optimization

Metrics inform decision-making and system optimization, enabling continuous improvement in fabric measurement accuracy and overall system performance. This section discusses how data-driven insights can lead to more effective decision-making and operational efficiency.

4.3 User satisfaction and acceptance

User feedback and satisfaction are critical in evaluating the system's impact. The AI-based system's performance was assessed based on user experiences, highlighting its effectiveness and ease of use in real-world textile manufacturing environments.

5. Result finding and analysis

In the results and analysis chapter, we have presented the findings of our study and experiment. This section mainly focuses on providing a detailed and objective account of the data trained, experiments conducted or observations made during g the implementation process. It contains the raw data, statistical analysis, and specifically performance metrics in a systematic and organized manner. The emphasis is on describing what is observed or measured and analyzing the results to draw meaningful insights and the best conclusions.

In this heading, the overall result obtained from this project is presented in a graphical and analyzed way. Lastly, we have shown the output result of the machine learning model in which the width of the fabric is shown on the screen to verify its accuracy and reliability.

5.1 Analysis of performance metrics

Mainly the performance metrics include the technical parameters that provide precise and accurate results. The recorded parameters are graphically described below (**Figure** 7):

- **train/box_loss**: This parameter is recorded during the machine learning model training. As the number of epochs increases, the box_loss value decreases. This parameter refers to the box showing during the detection of fabric, if the fabric gets moved from left to right the box also moves with it. It must be synchronized with the fabric movement.

- **train/obj_loss**: This parameter is also recorded during machine learning model training. As the number of epochs increases, the train/obj_loss value decreases. This parameter refers to the loss of detection of fabric, if the fabric is not detected by the model the loss will increase which will definitely reduce the accuracy.

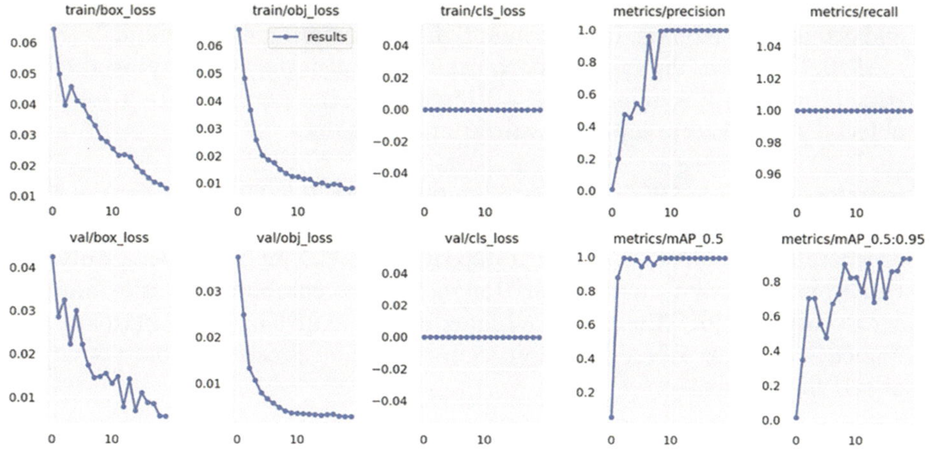

Figure 7.
Performance metrics of the model.

- **train/cls_loss**: This parameter is also recorded during machine learning model training. As the number of epochs increases, the train/cls_loss value decreases. This parameter refers to the classification loss, which measures the discrepancy between actual width and predicted width. The purpose of this metric is to evaluate how well the system is learning to classify fabric widths correctly.

- **val/box_loss**: This parameter is recorded after the model is trained. This parameter is working on the same specifications of train/box_loss.

- **val/obj_loss**: This parameter is also recorded after the model is trained. This parameter works on the same specifications as the train/obj_loss parameter.

- **val/cls_loss**: This parameter is also recorded after the model is trained. This parameter is working on the same specification of train/cls_loss.

- **metrics/precision**: This parameter mainly focuses on the precision of the model. It shows that as the number of epochs is increasing, the precision is also increased. There is a point where the precision is maximized and constant, this is because of the epoch's achievement, as the epochs are reaching to maximum the model is trying to give the best prediction with more precise values.

- **metrics/recall**: This parameter mainly focuses on the situation when the model recalls the training overall. If there are issues in accuracy, precision, and detection, the graph will show the ascending order toward the y-axis. In this scenario, the recall parameter is fixed at 1.0 because there are no such issues in accuracy and precision.

- **metrics/mAP_0.5**: The metric "metrics/mAP_0.5" refers to the mean average precision at an intersection over a union (IoU) threshold of 0.5. In object detection tasks, IoU measures the overlap between predicted bounding boxes and ground truth boxes. In the context of fabric width measurement, "metrics/mAP_0.5" assesses the system's performance in accurately identifying and localizing fabric edges. A higher mAP_0.5 indicates better performance in terms of object detection accuracy and localization precision.

- **metrics/mAP_0.5:0.95**: The metric "metrics/mAP_0.5:0.95" refers to the mean average precision across a range of intersection over union (IoU) thresholds, specifically from 0.5 to 0.95 with a step size of 0.05. The mean average precision (mAP) is a commonly used evaluation metric that combines precision and recall across different IoU thresholds. A higher mAP_0.5:0.95 indicates better performance in terms of object detection accuracy and localization across a wide range of IoU thresholds.

5.1.1 Precision-confidence curve

The precision-confidence curve in this model shows the relationship between precise and confidence scores for object detection. Basically, precision measures the accuracy of the system by correctly identifying the object in real time. Confidence scores show the stage of confidence associated with each predicted detection. The above curve helps to evaluate the trade-off between precision and confidence which allows us to select an appropriate confidence threshold that balances accuracy and reliability based on AI-based fabric width measurement system. The confidence set in this model is 0.7, hence the model will only show the width of fabric if it is confident 70% of its predicted value (**Figure 8**).

5.1.2 Precision-recall curve

Similarly, the precision-recall curve represents the relationship between precision and recall for object detection. As discussed above the precision shows the accuracy of the system to perfectly detect fabric in the running samples or real-time. Similarly, the recall factor measures the system's ability to correctly highlight the detection in running samples. Basically, the precision-recall curve provides insights into the system's performance by accurately detecting cloth edges and measuring the width. As the graph shows that the precision factor and recall factor are at maximum value, hence the model detects the fabric in real-time very perfectly (**Figure 9**).

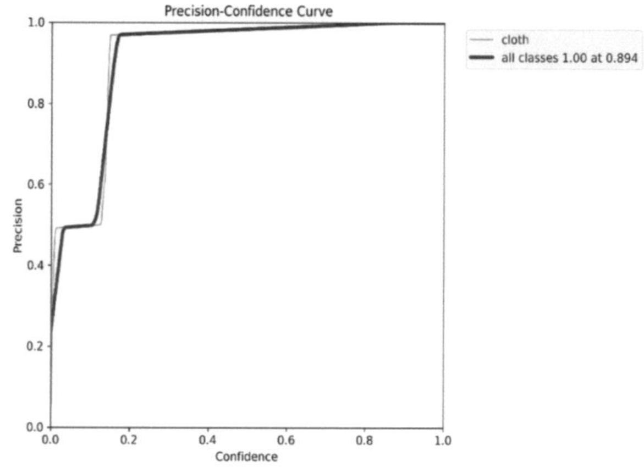

Figure 8.
Precision-confidence curve.

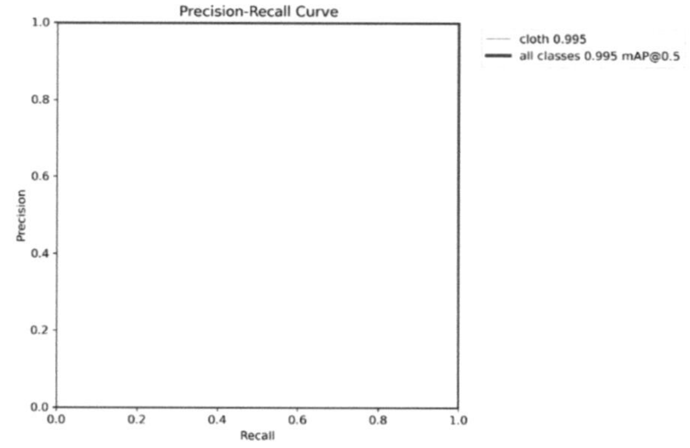

Figure 9.
Precision-recall curve.

5.2 Final outputs of the model

As we have analyzed successfully the model is running very smoothly. So, the output of the model must be verified for research satisfaction. Finally, the width of the fabric measured by the model is shown below (**Figures 10** and **11**).

The above output of the model is completely showing the objective of this project. The accuracy of the model detection is around 90%. Furthermore, the model has precisely detected the fabric passing through the inspection table. The width is also

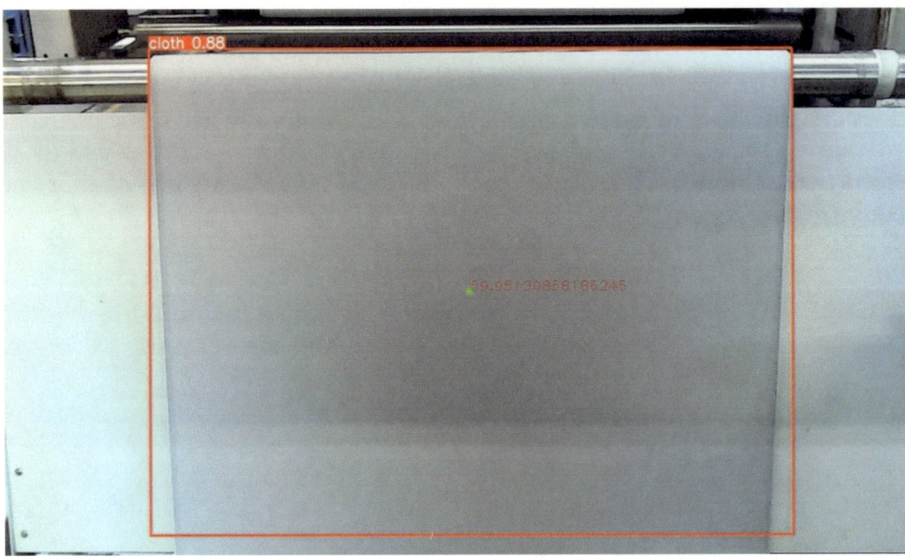

Figure 10.
Predicted width at 59.95inc. of fabric.

Figure 11.
Predicted width at 60.07inc. of fabric.

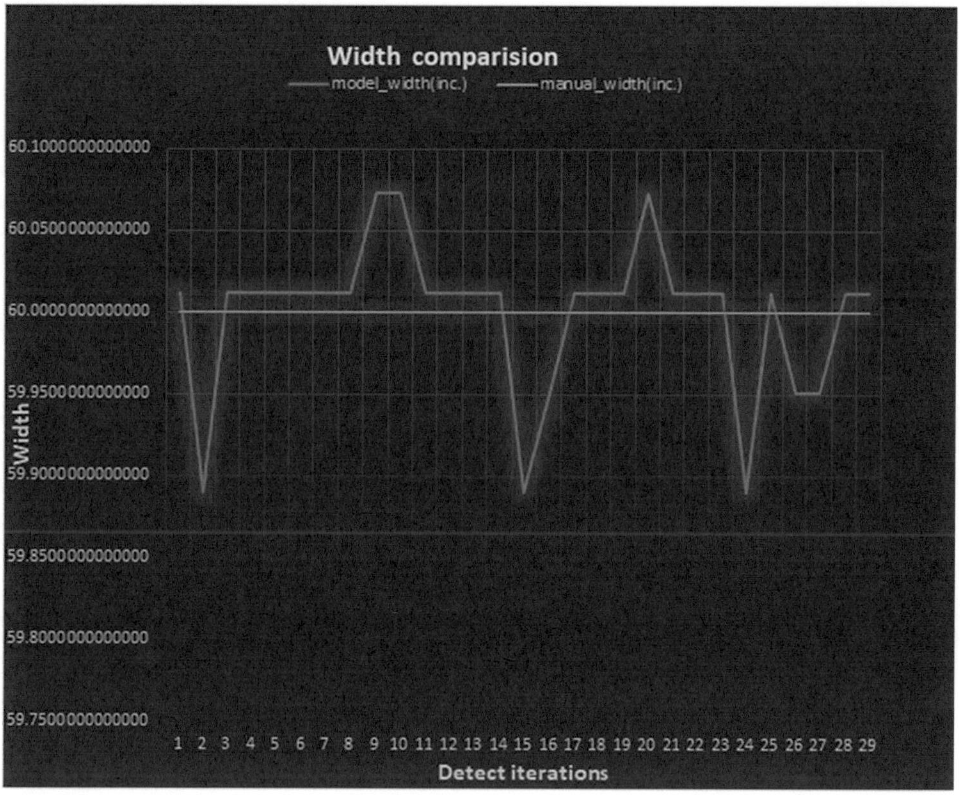

Figure 12.
Graphical representation of predicted width and manually measured width.

S.no	model_width(inc.)	manual_width(inc.)
1	60.0121728545343	60
2	59.8904443091905	60
3	60.0121728545343	60
4	60.0121728545343	60
5	60.0121728545343	60
6	60.0121728545343	60
7	60.0121728545343	60
8	60.0121728545343	60
9	60.0730371272063	60
10	60.0730371272063	60
11	60.0121728545343	60
12	60.0121728545343	60
13	60.0121728545343	60
14	60.0121728545343	60
15	59.8904443091905	60
16	59.9513085818624	60

S.no	model_width(inc.)	manual_width(inc.)
17	60.0121728545343	60
18	60.0121728545343	60
19	60.0121728545343	60
20	60.0730371272063	60
21	60.0121728545343	60
22	60.0121728545343	60
23	60.0121728545343	60
24	59.8904443091905	60
25	60.0121728545343	60
26	59.9513085818624	60
27	59.9513085818624	60
28	60.0121728545343	60
29	60.0121728545343	60

Table 1.
Analyzing the data of width of fabric.

showing which is updating in milliseconds due to the frame rate and precision of the trained machine learning model.

5.3 Comparison results between existing methods and this system

Lastly, we have compared the width of fabric by this AI-based fabric width measurement system with the manual measurement of fabric method which is currently applied on the machine. We have shown graphical representation and width analysis data below where the accuracy of this model could be seen easily without any discussion (**Figure 12** and **Table 1**).

6. Conclusion

In conclusion, we have summarized the important key points of experimental setup, performance metrics, and results and analysis which have given us more important new information about its efficiency and performance. The experimental setup, which included different technical aspects including data labeling, epoch size, and batch size, was extremely important in assessing the accuracy and reliability of the system. The system's precision, recall, and overall performance were assessed quantitatively using the performance metrics train/box_loss, train/obj_loss, train/cls_loss, and metrics/mAP_0.5:0.95. The evaluation of these metrics revealed the system's capacity for precise fabric width measurement as well as its potential for process and quality control improvement in the textile sector. Additional information was provided *via* the precision-confidence curve and precision-recall curve, which demonstrated the system's accuracy in fabric detection and breadth measuring. The model's excellent output, which provided accurate and immediate measurements of fabric width, confirmed the effective implementation of the AI-based system. The significance of the AI-based fabric width measurement system in increasing productivity and assuring high-quality fabric production processes is highlighted by these findings. The system is a useful tool for the textile industry because of its precision, dependability, and efficiency, which allows for better decision-making, process optimization, and user happiness. The fine output of the model is a noteworthy feature of the AI-based fabric width measurement method, in addition to the thorough analysis of the trial results. When detecting cloth as it moves through the inspection table, the model has shown outstanding accuracy and dependability. The result demonstrates the system's capacity to precisely recognize the cloth edges and offer in-the-moment width measurements. The fabric's width is altered in milliseconds, reflecting the machine learning model's high frame rate and accuracy. This high-quality output provides verifiable proof of the system's effectiveness and its potential to completely transform the way fabrics are measured in the textile industry. Manufacturers may rely on an artificial intelligence-based solution to provide precise and immediate measurements of fabric width, improving quality control, streamlining production, and increasing operational effectiveness.

Acknowledgements

Firstly, I want to thank you my group member Naresh Kumar who has implemented all the technical work of this project. We would like to acknowledge the amazing support and help of our internal advisor Mam Arham Iqbal and NED University of Engineering and Technology. It proved to be of great help having such strong support with us. Secondly, we would like to appreciate the limitless efforts of our external advisor Mr. Shehzad (Electrical Manager) at Naveena Exports. Lastly, Naveena Exports for funding our project and providing us with such a wonderful workplace environment to implement this project.

Notes/thanks/other declarations

We dedicated our project effort to Allah Almighty, who has provided us with strength throughout our lives. We would also want to dedicate this effort to our parents, friends, instructors, and all who have prayed for us, as we could not have completed it without their assistance and support. We would not have achieved such success in our lives if it had not been for them; without them, we would be nowhere. Thank you for your unwavering support of our studies. Without all of you, we would not have been able to achieve our objectives.

Author details

Muhammad Mirza Nouman Baig
NED University of Engineering and Technology, Karachi, Pakistan

*Address all correspondence to: baignouman8@gmail.com

References

[1] Li C, Li J, Li Y, He L, Fu X, Chen J. Fabric defect detection in textile manufacturing: A survey of the state of the art. Security and Communication Networks. 2021;**2021**:13, 9948808. DOI: 10.1155/2021/9948808

[2] Hossain MA. Artificial intelligence (AI) in textile processing. 17 Dec 2022. DOI: 10.13140/RG.2.2.20082.17600

[3] Preil D, Krapp M. Artificial intelligence-based inventory management: A Monte Carlo tree search approach. Annals of Operations Research. 2022;**308**:415-439. DOI: 10.1007/s10479-021-03935-2

[4] Ferro R, Cordeiro GA, Ordóñez REC, Beydoun G, Shukla N. An optimization tool for production planning: A case study in a textile industry. Applied Sciences. 2021;**11**:8312. DOI: 10.3390/11188312

[5] Sikka MP, Sarkar A, Garg S. Artificial intelligence (AI) in textile industry operational modernization. Research Journal of Textile and Apparel. 2024;**28**(1):67-83. DOI: 10.1108/RJTA-04-2021-0046

Chapter 5

Perspective Chapter: Textile Industry Challenges and Priority Actions to Mitigate Microplastic Fibre Pollution

Carmen Ka Man Chan, James Kar Hei Fang, Fei Bin and Chi Wai Kan

Abstract

The textile industry faces challenges caused by microplastic fibre (MPF) pollution. Urgent measures and interventions are needed to mitigate the release of MPFs throughout the textile lifecycle. Obstacles arise when implementing action plans that impede stakeholders from taking the appropriate steps. Standardised test methodologies to support the control of release are still in their infancy for application in the broader industry. The contribution of domestic and industrial wastewater to microfibre pollution is ambiguous, so considering natural fibres alongside synthetic alternatives has amplified the complexity. Instead of awaiting perfect solutions, the industry should prioritise implementing effective mitigation strategies without delay, including raising public awareness, fostering collaboration, integrating policies, improving wastewater treatment infrastructure, and supporting technological advancement. Selected sustainability initiatives that align with this agenda are utilised to generate insights and expedite actions.

Keywords: microplastic fibre, microfibre pollution, microfibre release, textile pollution, domestic laundry, textile effluent, sustainable textile

1. Introduction

MPFs are ubiquitous pollutants in various environments, primarily from textile fibres, released through multiple pathways. Limited stakeholder awareness and inadequate preventative measures due to the absence of global standards and policies hinder efforts to address this issue. While initial efforts focused on domestic laundry sources, growing attention now encompasses the entire fashion and textile value chain, highlighting significant industrial contributions. The industry has numerous opportunities to prevent MF release and entry into the environment, and MPFs can be repurposed as valuable raw materials or energy sources. Tackling this issue necessitates adopting a perspective that involves multiple stakeholders, taking a comprehensive approach, and sharing responsibilities among stakeholders throughout the

 IntechOpen

entire lifecycle of the garment industry, despite obstacles such as a lack of knowledge with inconsistent communication, lack of standardised methodology, absence of established policies, and undivided stakeholder responsibilities. Collaborative, multistakeholder approaches to address MPF pollution have been sluggish and require insights from successful industrial initiatives. This chapter investigates researcher and stakeholder concerns and analyses practical industry efforts to focus on priorities and accelerate progress.

1.1 Relationship of microplastics (MP) and microplastic fibres (MPF) pollution

Microplastic (MP) pollution has emerged as a critical environmental concern because of its widespread presence and negative impacts on aquatic ecosystems and human health [1–4]. MPs are defined as plastic particles smaller than 5 mm [5–7], including particles in the nanometre range (1 nm) [8]. MP pollution, a global environmental issue, was detected in all the tested water bodies [9]. It causes tissue damage, reduced growth, and mortality in aquatic organisms, particularly those at the base of the food chain [10]. Microplastic fibres (MPFs) are defined as fibrous or threadlike pieces of plastic with a length between 100 µm and 5 mm and a width of at least 1.5 orders of magnitude shorter [11–13], and are considered a subset of MPs with similar characteristics [14]. Typically, MPFs range from 11.9 to 17.7 µm in diameter and 5.0 to 7.8 mm for polyester, polyester/cotton blend, and acrylic fabrics and are 6 and 175 µm in diameter [15]. MPFs are often overlooked owing to methodological challenges [14, 16]. Shedding is a characteristic of textile materials, and its size varies depending on the fibre, yarn, and fabric parameters. Owing to mechanical and chemical reactions during production, washing, or use, fibres break into fragments and are released into air, terrestrial, or aquatic environments, posing threats to different ecosystems [17–19]. This phenomenon applies to fibres of natural origin, as well as MPFs from fossil fuels. Therefore, when referring to broader fibre fragmentation, microfibres (MF) generally include all textile fibres originating from natural, semi-synthetic, and synthetic sources.

Furthermore, microplastics have been found to influence soil quality and aggregation, which are crucial for ecosystem functions [20]. MPs are pervasive and have the potential to carry chemical pollutants, raising concerns for wildlife and human health and causing comprehensive research and international collaboration to address this multifaceted issue [3, 21]. In conclusion, the critical issue of MP is emphasised by their extensive dissemination, the severe ecological consequences they cause, and the potential hazards they pose to human well-being. The durability of these pollutants in the environment and their capacity to influence various tiers of the food chain underscores the need for intensified surveillance, regulatory measures, and mitigation strategies to safeguard ecosystems and public health [3, 4, 21]. Although MPFs are the most dominant form of MPs, most studies have primarily focused on MPs [22].

1.2 Scope and scale

As a result of economic and population growth, there has been an increase in the production of fibres worldwide from approximately 112 million tons in 2021 to 116 million tons in 2022. The primary reason for this increase is the production of virgin fossil-based synthetic fibres, which increased from 63 million tons to 67 million tons. Assuming this trend continues, it will reach 147 million tons by 2030. Synthetic materials typically include polyethylene terephthalate (PET), polyester, acrylic, polyamide, and polypropylene. Polyester is the most widely produced fibre globally, accounting for 54% of the total production by 2022 [23]. The demand for synthetic materials is projected to grow faster than natural fibres, primarily because of the need for emerging economies. Synthetic textile fibres became the most dominant fibre in the market, surpassing cotton in production volume in the mid-1990s. The forecasted value of total clothing sales in 2050, assuming business proceeds as usual, is expected to be more than three times that of today [24].

Attempting to decrease consumption or prohibit the production of synthetic fibres is not feasible unless alternative materials are accessible, owing to the benefits they provide. Synthetic fibres are popular due to their low cost and performance properties, such as strength, stain resistance, softness, and durability. They have relatively minimal carbon and water footprints, superior material performance, and recyclability compared to natural fibres. Apart from clothing, they are extensively applied in home textiles, decorations, toys, and other products.

MPFs have been found in aquatic and terrestrial ecosystems, extreme regions such as mountains and polar areas, and various food items consumed by humans [25]. The widespread presence of MPs highlights their importance, persistence, and ability to attract other pollutants, posing ecological impacts and potential human health risks [26]. The persistence and potential for MPF bioaccumulation necessitate continued research and mitigation efforts. MPFs are extensively distributed in freshwater, rivers, oceans, landfills, and air environments. Browne et al. [27] estimated that mechanisation was linked to population density at shoreline sites worldwide and indicated that 85% of MPs were MFs. Although research on MP pollution is still in its early stages, much remains to be discovered. With limitations in sampling and methodology and significant variations in the quantities reported, the findings still indicate a greater abundance of MFs than other types of MPs [15, 28].

The persistence and potential for MPF bioaccumulation necessitate continued research and mitigation efforts. MPFs are extensively distributed in different environments, such as freshwater, rivers, oceans, landfills, and air. Browne et al. [27] estimated that mechanisation was linked to population density at shoreline sites worldwide and indicated that 85% of MPs were MFs. Although research on MP pollution is still in its early stages, much remains to be discovered. With limitations in sampling and methodology and significant variations in the quantities reported, the findings still indicate a greater abundance of MFs than other types of MPs.

Boucher and Friot [29] estimated that synthetic textiles accounted for 35% of the global release of MPs into oceans, with 25% emerging from the wastewater pathway. By combining the releases of MPFs shown by existing domestic laundry studies, the Ellen MacArthur Foundation [24] estimated that over 22 million tons of MPFs will be added to the ocean between 2015 and 2050 if demand continues to follow the current pattern. Approximately 2% of global production is projected to be lost to the ocean.

1.3 Cause of concern

As a subset of MPs, MPFs are anticipated to exhibit comparable characteristics. MPs are resistant to natural degradation and have not been degraded for decades. Their continuous release results in their inevitable increase and accumulation in the environment. Increasing evidence has suggested that MFs are the most prevalent type of secondary MP debris in marine environments [27, 30]. The most prominent and widely reported sources of MFs are shedding during the domestic laundry and manufacturing processes [14, 29, 31–33].

Mechanical and chemical stresses are known to cause MF release during domestic washing or industrial wet processing of fabrics during production and usage [34]. In the entire product lifecycle, fibre spinning into yarn, knitting or weaving into the fabric, and sewing panels into products, fibres are released throughout the entire manufacturing process to different extents to the weakening of fibres and breaking into fragments [35–37]. Suction devices collect a significant quantity of released fibres at production, disposed of as solid waste or as effluent and sludge after the fabrics are treated in wastewater treatment plants (WWTPs) of wet processing mills.

Although WWTPs are thought to be 95–99% effective, these plants are not explicitly designed for MP retention, which results in MP retention and release into the aquatic environment [38–41]. As MPs/MFs can bypass WWTPs and have enormous discharge volumes, there is sound evidence that WWTPs are significant sinks for MP pollution. Moreover, it is crucial to note that MFs are the most abundant type of MP found in many WWTPs [15, 40, 42, 43].

Longer physical characteristics and fibrous MFs appear to have more significant environmental consequences than regular-shaped MP particles. This fibrous shape increases the tendency for entanglement in the digestive tract, leading to the blockage or abrasion of internal organs, a higher chance of compromised growth, adverse effects on reproduction, and starvation [44–46]. The larger surface area of fibres is also significant as it potentially allows greater sorption of harmful compounds and higher retention in the gut, consequently allowing more time for the leakage of plastic additives. Cole [47] stated that fibrous MPs or MFs might present a more significant risk than spherical MPs. Tiny and fibrous materials at the microscopic scale have the potential to induce cancer and fibrosis, whereas particles with equivalent particulate content are typically non-threatening.

There are only a few studies on the bioaccumulation of plastics and the associated chemical effects of absorbed substances such as carcinogenic persistent organic pollutants (POPs) [48–50]. The importance of MPs, specifically MFs, as vectors for transporting harmful chemicals relative to alternative pathways remains a mystery, with neither their quantities nor sources identified [15, 51]. However, the ingestion of organisms has been demonstrated to have the potential to introduce toxic compounds into higher trophic-level species, causing adverse effects as they are transferred to the food chain and potentially reach humans [43, 49, 52]. MFs surround us because they exist everywhere, but little is known about their long-term impacts on human health and the natural environment. More discussions are emerging regarding their fate in the atmosphere we breathe, the particles we inhale at home, the water we consume, and other aspects of our beings and environment [53, 54].

According to industry forecasts [55], synthetic textile production is expected to grow remarkably, and MPF is magnified and more challenging to resolve. A systematic approach is vital to address and manage the problem throughout the entire textile value chain, up to end-of-life disposal. This emerging issue is closely associated with the United Nations' Sustainability Development Goals (UNSDG). Regarding life in seawater (No. 9) and responsible production and consumption (No. 12), global efforts should be made to resolve them collectively without delay.

2. Literature review

2.1 Sources and microplastic fibres

Based on Textile Exchange [55] findings, synthetic fibres and materials comprise 64% of all fibres and materials produced globally across sectors, including fashion, apparel, and textiles. Polyesters are the most common in the broad synthetic category, accounting for 54% of global fibre production in 2022. Global fibre production demand has significantly grown from 8.4 kg per person in 1975 to 14.3 kg per person in 2021, with less than 0.6% recycled from pre- or post-consumer textiles. If the business persists with the status quo, the definite growth trajectory will be 34% from 2020 to 2030 [23]. The reason for the growth and shift in fossil fuel-based plastic fibres is that they cost less and provide better functional performance than natural fibres such as cotton [11, 56].

The most prominent and widely reported sources of MPFs are shedding during domestic laundry and manufacturing processes [14, 57]. For an average wash load of 6 kg, over 700,000 fibres could be released per wash and discharged into the aquatic environment via WWTPs. Although WWTPs are reported to be 95–99% effective, they are not explicitly designed for MP/MPF retention; therefore, they can bypass WWTPs. Owing to their enormous discharge volumes, there is compelling evidence that WWTPs are significant sinks for MP pollution [43]. Other sources of wastewater from WWTPs may include solid waste or sludge disposal. Sludge, a byproduct of WWTPs commonly applied to agricultural land as a fertiliser, has been found to contain MFs. Tao et al. [58] estimated that during a 15 min drying process, over 90,000 microfibres could be released from 1 kg of polyester and cotton textiles. The release of MFs from dried clothes can represent a source of airborne MP pollution [58–60]. These byproducts, which appear as airborne textile fibres, can also cause persistent terrestrial contamination [46, 61–63]. These pathways are involved in the release of MFs into the environment.

The ecosystem impacts of MPs/MFs have the potential to occur through physical, chemical, and biological pathways. Woodall et al. [30] initially reported ingestion of MFs by deep-sea organisms in natural settings. Its fibrous and lengthy nature causes concerns regarding the buildup of MF in the environment. MFs tend to entangle and block the digestive tract, leading to starvation and impaired growth of microorganisms [46]. The longer-term impacts include the capacity to absorb harmful chemical substances from fibres associated with dyes or additives used in textile manufacturing. Several studies have reported that MFs can transfer contaminants. The ingestion of MFs by various aquatic species, including turtles, seabirds, fish, and lobsters, is associated with reduced feeding and reproductive abilities.

Furthermore, owing to the large surface area of MPs/MFs, toxic compounds can be introduced into humans by eating higher trophic-level species because these compounds are transferred along the food chain [64]. Ragusa et al. [65] identified MPs in the human placentas. Increasing evidence of MPFs in the atmosphere has recently been reported. Flying inhaled MFs can be deposited in the lung tissue [66] and may lead to tumours. According to Cole [47], nanoscopic and microscopic fibrous materials can be carcinogenic and fibrotic, whereas particles with the same content are comparatively benign. The fact that MFs are increasing and accumulating toxins that are dangerous to marine life and potentially humans will only magnify and become more challenging to resolve if further action is delayed.

It is worth mentioning that, at present, many studies on MPF pollution have only focused on synthetic fibres and excluded other natural semi-plastics, such as man-made cellulosic-based fibres, and their potential impact. Controversy surrounds the discussion of whether they should all be included. Fishing nets, ropes, and gears are potential sources of MPF pollution. The primary concern regarding the use or abandonment of fishing gear is the source of secondary MP pollution. It is perceived that ropes discarded following their use in fishing activities and marine transportation are the primary sources of MFs following weathering and material breakdown over time [13, 67, 68]. This discussion has been influenced by studies such as Miller et al. [69] and Dris et al. [70, 71], who identified both synthetic and non-synthetic fibres in freshwater and even in the gastrointestinal tracts of fish [48].

2.2 Microplastic fibre release from domestic laundry

2.2.1 Washing parameters

Several studies [25, 37, 72–74] reviewed MF shedding during the usage phase. Different conditions like detergents, temperatures and machines are used to wash fabrics and garments. The laundry parameters outlined in **Table 1** unequivocally determined that laundry settings play a crucial role in minimising MF release.

One of the early studies by Browne et al. [27] reported that >1900 fibres were released per garment per wash. This was followed by Pirc et al. [75], who first stated that tumble drying releases 3.5 times more MPFs. Several studies have reviewed the shedding of MPFs during the usage phase. Fabrics and garments are washed under different conditions, such as detergents, temperatures, and machines. Variations in sampling, test methodologies, and reporting units employed in numerous studies make comparing results difficult.

Front-load washing machines minimise MF release owing to milder mechanical action, potentially up to seven times less than other machines [32]. Studies have indicated a decreasing trend in fibre release with successive washes [59, 78]. Washing load significantly influences fibre release, while textile parameters are also critical. Although compact structures, higher twists, and long filaments reduce MF release, translating this into a textile design is complex without detailed reports and comparisons across production processes. Rathinamoorthy and Balasaraswathi [25] found that waste volume was the most influential laundry parameter for MF release, followed by washing duration, mechanical agitation, and temperature. Thus, maintaining a full load minimises the water-to-fabric ratio and effectively reduces fibre shedding during washing, which is consistent with the results of Kelly et al. [77].

Both McIlwraith et al. [90] and Napper et al. [1, 91] proved that mitigation devices had some impact on reducing fibre release during washing processes. The Lint LUV-R filter was the best performer, with a retention rate of approximately 80%. The number of remaining devices is significantly lower (approximately 30%). Herweyers et al. [92] confirmed that further education is required if these devices are introduced to consumers. They must be user-friendly and easy to understand, as well as the impact that consumers can make in securing long-term behavioural change.

From the evidence, the first wash releases most of the MF and reaches a plateau after several washes [77, 78, 93]. According to Mahbub and Shams [81], the release

Washing conditions	Reduce MF release (positive impact)	Increase MF release (negative impact)	References	Details (MF release)
Machine type	Front Load	Top Load	[32]	The top load releases 7 times more than the front load
Continue subsequent washes	+	N.A.	[75]	Successive washes decrease emission
	+	N.A.	[76]	Reduce and typically stable from the 4th and 5th cycle
	+	N.A.	[77]	Reduce and typically less after 4th cycles
	+	N.A.	[78]	Reduce after the peak at 3rd cycle
	+	N.A.	[59]	Reduce and stabilise from 5th cycle
	+	N.A.	[79]	Reduce
	+	N.A.	[80]	Reduce significantly from 5th cycle
	+	N.A.	[81]	Reduce and stabilise at the 7th cycle
	+	N.A.	[82]	Reduce and stabilise from 3rd cycle
Liquor (Water-volume-to-fabric) ratio increase/ washing load decrease	—	+	[77]	Increase as the most influential factor
	—	+	[83]	Increase by 5 times
	—	+	[82]	Increase 2 times with liquor ratio increase from 7.5:1 L/m^2 to 15:1 L/m^2
	—	+	[84]	Increase with smaller loads (higher liquor ratio)
	—	+	[25]	Increase as the most influential factor
Washing temperature increase	—	+	[15]	Increase with temperature
	No impact	No impact	[85]	No significant effect
	No impact	No impact	[86]	No significant effect
	—	+	[87]	Increase with temperature
	—	+	[77]	No significant effect between 15 and 30°C and increase at 60°C
	—	+	[88]	Increase with temperature
	—	+	[89]	Increase with temperature
	—	+	[73]	Increase with temperature
	—	+	[81]	1.8 times more if the temperature increased from 20 to 40°C
	—	+	[74]	Increase with temperature between 25 and 60°C
	+	—	[82]	Reduced the microfibre release by 26% with temperature increase from 40 to 90°C
Washing and drying time increase	No impact	No impact	[77]	No impact if the increase from 15 to 60 min
	—	+	[73]	Increase with duration and spin speed
	—	+	[81]	Increase if it is longer from 30 to 60 min
	No impact	No impact	[82]	No impact on washing time 30 min and 60 min
	No impact	No impact	[74]	No impact on washing time 5-20 min

Washing conditions	Reduce MF release (positive impact)	Increase MF release (negative impact)	References	Details (MF release)
Adding detergent and softener	+	—	[75]	Reduce (both detergent and softener)
	+	—	[86]	Reduce (softener only)
	—	+	[87]	Increase (detergent only)
	—	+	[88]	Increase with detergent
	No impact	No impact	[77]	No effect (detergent only)
	+	—	[78]	Reduce (detergent only)
	No impact	No impact	[84]	No significant impact (both detergent and softener)
	—	+	[73]	Increase (detergent and conditioner), powder detergent>liquid detergent
	+	—	[25]	Reduce (softener only)
	—	+	[81]	Increase (detergent only)
	—	+	[74]	Increase with detergent (both handwash and machine wash)
Agitation (Stimulated by steel balls)	—	+	[82]	Increase

Table 1.
Summary of effect on MF release from domestic laundry test in various washing conditions from selected papers.

could be reduced to 67% from the first to the seventh wash cycle. As durability becomes essential to reduce waste, MF release life cycle analysis becomes more crucial to a more holistic approach for estimating the overall impact of MF pollution.

2.2.2 Textile parameters

Hazlehurst et al. [82] indicated that while no single fabric characteristic predominantly influences MF release, fabric variables may have a more significant impact than laundering conditions. This suggests a significant potential for MF reduction through fabric variables. However, the concluding fabric parameters are complex because of the numerous possible combinations that can number in thousands. Modifying textile surfaces, such as mechanical brushing or adding chemical finishes to enhance the handfeel, may considerably affect MF release. Isolating the effects of individual production parameters is challenging because it is difficult to compare a single parameter while controlling for others. **Table 2** summarises the impacts of various textile parameters on MF release during domestic laundry, as reported in the selected studies.

Carney Almroth et al. [101] reported that polyester fleece fabrics with a more open structure shed most severely, about 84 times more than other knit structures. Weft knits have a higher MF release than warp knits because warp knits have a tighter and more compact structure than weft knit fabrics, which is similar to the findings of Carney Almroth et al. [101] and De Falco et al. [98]. Vassilenko et al. [102] suggested that mechanically treated polyester samples, dominated by fleeces

Textile Parameters	Reduce MF release (positive impact)	Increase MF release (negative impact)	References	Details (MF release)
Composition	Polyester	Blends (Polyester Cotton)	[94]	Polyester Cotton blend> Polyester
	Polyester>Acetate	Polyamide>Acetate	[88]	Polyamide>Acetate>Polyester
	Synthetic	Cotton	[78]	Cotton>Synthetic (Acrylic>Polyamide>Polyester)
	Virgin polyester	Recycled polyester	[93]	Recycled polyester (MF shorter length) > virgin polyester
	Polyester	Cotton/Cotton % in blend	[95]	Cotton>Polyester, Increase with higher % of cotton for blends
	Polyester	Cotton	[80]	Cotton>Polyester, increase with higher % of cotton for blends
	N.A.	Increase elastane	[96]	Increase with % of elastane
	TC (Polyester rich) < CVC (Cotton rich)	Cotton>polyester	[72]	CVC (cotton rich blend)) > TC (polyester rich blend) > Polyester>Cotton
Fibre length increase	+	—	[97]	Spun staple>filament
	+	—	[94]	Staple>filament
	+	—	[88]	Spun staple>filament
	+	—	[98]	Spun staple>filament
	+	—	[78]	Increase release with shorter irregular fibres
	+	—	[93]	Spun staple>filament
	+	—	[80]	Reduce from staple to textured filament.
	No impact	No impact	[82]	No significant between staple and filament
Spinning method	Air-jet	Ring	[99]	Ring>rotor or air-jet
	Air-jet	Ring	[100]	ring-spun>vortex-spun (air-jet)
Higher yarn twist	+	—	[101]	Reduce with higher twist
	+	—	[94]	Reduce with higher twist compact woven filament>staple lower twist knit
	+	—	[98]	Reduce with a higher twist.
	+	—	[79]	Spun > non-twist filament> hard twist filament

Textile Parameters	Reduce MF release (positive impact)	Increase MF release (negative impact)	References	Details (MF release)
Finer yarn count	+	—	[25]	Increase with coarser yarn
	+	—	[100]	Reduce with finer yarn count from 30tex to 15tex
Denser(Compact) structure	+	—	[101]	Increase with loose construction
	No impact (Density)	No impact (Density)	[86]	No significant difference on fabric density
	Woven	Knitted	[94]	Reduce with tighter woven>loose knitted
	+	—	[88]	Reduce with tighter/compact structure
	Woven	Knitted	[102]	Fleece and jersey > woven filament
	N.A.	Hairiness	[73]	Increase with hairiness
	+	—	[79]	Reduce with tighter or higher density
	+	—	[103]	Reduce as interlacing coefficient and weft density increase
	+	—	[25]	Reduce with stitch density, tightness, and loop length
	N.A.	+ (thicker)	[104]	Increase with fabric density, thickness
	No impact (form)	No impact (form)	[100]	No significant difference on hank form or knitted fabric form
Finishing	Pectin-based	N.A.	[105]	Reduce with a Pectin-based finish
	N.A.	Surface process	[99]	The processed surface can be 5 times more
	Mechanical finish (peaching)	N.A.	[82]	Reduce by mechanical finish (peaching)
Cutting	Laser-cut	Scissor-cut	[99]	Scissor-cut 3–31 times higher than laser-cut

Table 2.
Summary of effect on MF release from domestic laundry tests using different textile parameters from selected papers.

and jerseys, released six times more microfibres than nylon samples with woven structures and filamentous yarns. It is surprising to find that woven twill has the highest release compared to all knitted fabrics, which is also contradictory to other findings which believe woven fabrics are smoother and have a looser structure than knitted fabrics [82, 106].

Klinkhammer et al. [107] investigated various functional polyester garments and discovered that mechanically raised textiles, like fleece, rarely exhibited higher emission values than non-raised ones. They found that microfibre (MF) release could be more significantly influenced by altering washing parameters, particularly with an increased washing machine load and reduced washing time. Similarly, Cai et al. [99, 108] observed that processed surface textiles (e.g. fleece, plain brushed, and microfibre) released fibres at rates up to five times higher than those of unprocessed surfaces, indicating that abrasive friction during production is highly dependent on brushing effects. However, this does not imply that the release rates were higher for all fabric structures. In contrast, Hazlehurst et al. [82] found that mechanical finishes such as peaching can reduce MF release.

Özkan and Gündoğdu [93] found that recycled polyesters may release more microfibres (MFs) than virgin polyesters. Choi et al. [109] indicated that blended fabrics release more MFs than pure fabrics. Rathinamoorthy et al. [96] noted that increasing elastane content raises overall fibre emission, proportional to the elastane percentage. For example, a fabric with 2% elastane exhibited a 45% increase in MF release after the fifth wash. Julapong et al. [72] observed that fabrics made from natural, man-made, and mixed materials released more MFs than those made solely from synthetic materials. By all means, due to the inability to compare full specifications and their relationships, it is impossible to determine which factor has a more significant impact. Without direct comparison, these relationships can only serve as guidance, and measurement or testing is required to understand the actual MF release from a particular fabric quality.

2.2.3 MF release at sewing

MF release data are often undervalued and ignored when capturing data; little attention has been paid to MF generated during sewing. From these data, it is well known that what is measured is performed. Therefore, it is critical to include these data, although some can be released via a different pathway before washing. Fontana et al. [110] and Rathinamoorthy and Balasaraswathi [111] demonstrated that sewing directly affects MF release. A later study confirmed that the proper selection of stitch density and seam type can reduce the number of MF releases by up to 64.6%. They must be captured separately to address issues with the matching solutions effectively. Rathinamoorthy and Balasaraswathi [112] showed that sewing thread contributed approximately 1.09% of the total MF emissions during laundry and that the fibre of the sewing thread weakened after sewing. Finer count, filament, and twisted threads release less MF than coarser, spun, and untwisted threads.

Considering the negative impacts of sewing and sewing threads on microfibre release, it is essential to investigate alternative methods. Although using sewing threads specifically designed to reduce fragmentation is advantageous, it is also important to consider minimising the number of seams in garment construction and adopting seamless manufacturing techniques. However, comprehensive research is currently lacking to fully understand the potential effects of these strategies.

2.3 Test methodology in MPF release quantification

There is no standardised method for analysing MPFs [113, 114], particularly across research and industry communities. Measuring and identifying MPs is difficult due to inconsistent sampling, extraction, and quantification methods, resulting in challenges in comparing data and estimating their prevalence [16, 115]. As a result, the reported MPF concentrations among these studies were conflicting, varied, and divided without a standardised concentration reference. Many studies have been conducted on oversimplified tones with missing critical details. The concentrations of MPs/MPFs are relatively complex to compare because of the different units used [116]. Conversion efforts may lead to biased or misleading results because of necessary assumptions [117].

Recently, a few commercialised standards have been published for quantifying MPFs from domestic washing. These are published by the Microfiber Consortium (TMC) [118], AATCC [119], and ISO [120] and are released at the latest time. These are remarkably similar and consistent, as shown in **Table 3**; therefore, the results are projected to be comparable. From the perspective of procedure comparison, a minor difference between the TMC method is the requirement for eight specimens but no blank test. The AATCC and ISO standards also reduced the sample specimen quantity from eight in TMC to four. The AATCC provides two options for pretreatment and drying: detergent and wash temperature. The sample specimen size was larger to accommodate a wider seam allowance, allowing thicker fabrics to be sewn more easily. However, MPF testing is not widely used. As the early released industry bodies, The Microfibre Consortium [121] reported that about 850 fabric qualities were tested by their members in the last few years using their method and stored in their data portal. Potentially, AATCC adoption will be comparable or more because it is internationally recognised and does not require data to be reported to the TMC and become their signatory. Whereas ISO is relatively new, its adoption is expected to be very low.

Without addressing the calls to include counting as a more precise option to quantify MPFs [111], these methods capture only the weight and % weight of MPF released at the first wash. Compared to academic research, which goes deeper and includes counting and multiple washes, commercial testing provides limited information for textile producers and manufacturers to develop action plans; it is more appropriate for monitoring the aggregated reduction performance of the industry.

2.4 Pathways of microplastic fibres

2.4.1 Effluent from wastewater treatment plant

Wastewater effluent is a well-known source of MPFs, which are released into the aquatic environment. It is accepted that laundering fabrics leads to abrasion and wear of textiles, leading to fibre shedding and their subsequent release into sewage effluent from washing machines. Browne et al. [27] reported that the proportions of polyester and acrylic fibres in sewage effluents resemble those of MPs found in contaminated sediments worldwide. This indicates that some MFs in the marine environment originate from textile washing with WWTPs as pathways.

WWTPs have recently attracted attention as pathways for MP release, with an exponentially growing number of related publications in the last few years [122]. Mintenig et al. [123] suggested that WWTPs could be a sink and a source of MPs and thus play an essential role in MP pollution. There is strong evidence that MPs/MPFs

Standard	TMC Method-2019	AATCC TM212–2021	AN ISO 4484-1:2023
Description	Quantification of fibre release from fabrics because of domestic laundering	Test Method for Fibre Fragment Release During Home Laundering	Microplastics from Textile Source - Determination of material loss from fabrics during washing
Scope	All textile materials	All textile materials	Synthetic and natural fibres
Sample Specimens	8	4	4
Resultant Dimensions	100 mm ± 10 mm x240mm ± 10 mm	100 ± 10 mm x 240± 10 mm	100 ± 10 mm x 240± 10 mm
Original Dimensions	150 mm ± 10 mm x290mm ± 10 m	200 mm ± 10 mm x340mm ± 10 m	150 mm ± 10 mm x290mm ± 10 m
Pretreatment	Dry at 50 °C 4 hr	Option A: 21 ± 2°C, 65 ± 5% RH 4 hr. *Option B: 70 ± 2°C 4 hr. (desiccator 30 min)	50 ± 3°C dry to constant mass, cool in a desiccator
Liquor	360 ml, grade 3 water, 50 steel balls	360 ml, grade 3 water, 50 steel balls	360 ml, grade 3, 50 steel balls
Detergent	Nil	Option A: 0.25% Option B: Water only	Nil
Wash Conditions	40 °C for 45 min, with a rotation speed of 40 ± 2/min	Wash temperature as the label for 45 min, rotation speed of 40 ± 2/ min	40 ± 3°C for 45 ± 1 min, rotation speed of 40 ± 2/min
Blank Test	Not required	Required, every 4 specimens	Required, every 4 specimens
Filter	1.6 μm pore size, 47 mm diameter Dry at 50 °C 4 hr	1.6 μm pore size, 47 mm diameter Rinse and dry as Pretreatment	1.6 μm pore size, 47 mm diameter, Dry at 50 ± 3 °C 4 hr
Balance Precision	0.0001 g	0.0001 g	0.1 mg
Calculation	Fibre release (%) = 100x(W2-W1)/S	Fibre fragment release (g) = W2-W1 Fibre fragment release (%) = 100x(W2-W1)/S	Fibre fragment release (g) = M2-M1 Fibre fragment release (%) =100x(M2-M1)/S
Reporting	Average mass of fibre fragment release (significant digits 3) Average % of fibre fragment release	Average mass of fibre fragment release and standard deviation Average % fibre fragment release and standard deviation	Mean of mass loss Mass loss % of the original specimen

Table 3.
Comparison of commercial microplastic fibre release testing methods. Consolidated from TMC Method-2019, AATCC TM212–2021 and ISO/DIS 4484–1 standards.

can easily bypass WWTP filtration and other solid separation processes that have not been designed for this purpose [40, 41, 124]. Magnusson and Norén [39] and Talvitie et al. [125, 126] demonstrated that the supply of MPs from WWTP effluents to the aquatic environment may be substantial because of the enormous volume discharged daily. In the study by Conley et al. [124], the MPF removal efficiency was reported at 80.2–97.2%, which is relatively high but still significantly less than the total MPs. Talvitie et al. [125, 126] suggested that advanced wastewater treatment (e.g. a membrane bioreactor) can improve the removal efficiency of small-sized MPs (<100 μm).

This was further supported by Ziajahromi et al. [41], who found that fibres were the dominant MPs detected in most effluent samples and were not completely removed even after advanced treatment processes.

Therefore, WWTPs in developed countries are considered to be more efficient. For example, Mintenig et al. [123] reported a removal efficiency of 98% for MFs after advanced filtration in a German WWTP. WWTPs usually have lower standards in developing economies because of inadequate sewage infrastructure. In 2014, China alone accounted for 69% of all polyester fibre production globally, with the combined output of China, India, and Southeast Asia representing over 80% of the global total [14]. Nevertheless, developing countries produce and consume more synthetic textiles (62.7%) than developed countries (48.2%) [29]. Developing countries tend not to have commonly available tertiary treatment standards, which is of great concern.

Xu et al. [127] measured MPFs directly discharged from textile mills to an industrial WWTP that collected effluents from mills in the region for treatment before their discharge into the aquatic environment. This is a typical textile industrial Central Effluent Treatment Plant (CETP) with a daily treatment capacity of 30,000 tons in the region where this study was conducted. This plant receives production wastewater from 33 printing and dyeing mills in a textile industrial park, accounting for approximately 95% of the influent. The average abundance of MFs, both natural and synthetic, was reported as 334.1 (±24.3) items/L in the influent before being reduced to 16.3 (±1.2) items/L in the final effluent, with a retention efficiency of 95.1%.

In China, when comparing industrial textile effluents from Xu et al. [127] and municipal effluents from Yang et al. [88, 128] and Lv et al. [129], there were 28–1310 times more MFs discharged directly in concentrated effluent to aquatic environments from industrial sources than from municipal WWTPs. The scale of textile effluents is vast, one of the highest water-use industries globally [130], and it is considered the second largest consumer and polluter of clean water [24]. Once discharged, no practical solution exists to remove it from freshwater and the ocean. It is essential to avoid discharging MFs into aquatic environments because there is no alternative way to remediate them. Therefore, it is imperative to take immediate action to control wastewater release. The magnitude of this pathway suggests that it is of primary importance and a top priority. Regulators should be informed and educated without any delay.

2.4.2 Sewage sludge (terrestrial)

Upon further examination of the studies by Habib et al. [131] and Zubris and Richards [132], it has been shown that synthetic fibres were reported as an indicator of municipal sewage sludge used in soils, indicating fibre presence in wastewater and their diverging pathways. MFs within the sludge material cannot biodegrade quickly in nature and persist for more than 15 years in the soil. Such pollutants in the soil can be retained in terrestrial environments or eventually enter marine environments through runoff.

MFs can also be discharged into terrestrial soils by applying sludge as a fertiliser to farmlands [32] because of their nutritional and organic content. Since most of the MFs found in WWTPs are kept in sludge [125, 126], this pathway for MFs has drawn considerable attention because it captures fibres retained by highly efficient WWTPs and transfers them to the terrestrial environment. However, there is limited evidence of the impact of MFs on soil or land.

2.4.3 Seawater and marine sediment

MFs are found in seawater and sediments across all global marine habitats. Gago et al. [133] reported that the most common fibre length of surface water is 500–1000 μm, reaching 4750 μm. In sub-surface waters, the length is 1–5 mm, and in sediments, it ranges between 800 and 1000 μm, which is consistent with findings from studies on WWTPs and domestic laundry. MFs and MPFs are inevitable byproducts of modern lifestyles and can reach oceans, rivers, and lakes directly, bypassing WWTPs [31].

A Great Lakes study indicated that the estimated MPFs in aquatic environments are significantly higher than the concentrations in wastewater effluents [134]. Similarly, Klein et al. [135] identified discrepancies between effluent fibres and overall environmental counts in river shore sediments. Additional research has shown that deep-sea sediments serve as major sinks for MFs [136]. MPs also enter marine environments through storm sewers, winds, currents [137], and runoff [61, 138]. MP debris is degraded via sea recycling ports and landfills facilitated by adverse weather conditions. MFs from fishing gear and lines may be the primary marine sources [133].

2.4.4 Atmospheric fallout

Besides the aquatic environment, Dris et al. [139] first reported the concentration of 1.0 and 60.0 fibres/m^3 and 0.3 and 1.5 fibres/m^3 in indoor and outdoor air, respectively. Most were natural or cellulosic, as short staple fibres are more prone to breakage and fallout during use. Research on prolonged ultraviolet radiation exposure and ambient temperature variations has shown that textiles deteriorate and emit MFs into the atmosphere [140].

Research suggests that laundering is the most significant cause of textile fibre shedding, but this may only apply to certain products. Non-laundered fabrics such as flags and sails also deteriorate over time, leading to fibre loss [31]. The fibres were readily visible on various indoor surfaces. Recent studies have discussed the inhaled air quality, household dust, drinking water, and other environmental components. Liu et al. [140] and Zhang et al. [141] indicated that textile garments and soft furnishings are likely the primary sources of airborne MPs indoors, with over 60% being MFs, potentially a significant source of MP pollution.

Textiles are a major source of airborne microplastics, predominantly in a fibrous form. Microplastics ≤10 μm, particularly those <5 μm, are more likely to be inhaled and pose significant health risks [142]. Studies [143–145] have revealed that most airborne fibres are smaller than 1000 μm, creating potential inhalation hazards. Recent research indicates that atmospheric microfibres can cause respiratory issues and have been detected in human lung tissues [66].

The MF concentrations differed markedly between indoor and outdoor environments across regions, as documented by Refs. [86, 140, 144, 145]. Clothing and textile furniture are the primary sources [146]. Variations in indoor environments are primarily attributed to differences in lifestyle and textile product usage. Outdoor distribution is significantly influenced by wind, airflow, and factors such as consumption habits, socioeconomic status, traffic, and urbanisation [147].

Nevertheless, the airborne MF phenomenon requires strict contamination control from sampling to the point of the experiment. This is particularly important because the potential presence of airborne MFs, such as from researchers' clothing or other textile materials, can significantly affect experimental outcomes. Therefore, it is essential to implement effective control procedures to minimise contamination.

2.5 Ecological impacts

2.5.1 Environmental risk

The ecosystem impact of MPs can arise through physical, chemical, and biological pathways. Woodall et al. [30] first reported ingestion of MF by deep-sea organisms under natural conditions. However, the long-term effects of different polymer types and their abundances [148] remain largely unexplored. These impacts include the ability to absorb harmful chemicals from fibres related to dyes or textile additives and the potential for MFs to accumulate environmental toxins. Studies have documented the transfer of contaminants, such as plasticisers [149], dyes [150], polycyclic aromatic hydrocarbons, and polychlorinated biphenyls [151].

Fibrous MPs may significantly threaten marine life more than spherical MPs. Microscopic fibrous materials can induce carcinogenesis and fibrosis; however, particulate forms of the same material are less harmful [138]. Ingestion of MFs by aquatic species can lead to entanglement, stunted growth, and decreased feeding and reproductive abilities [148, 152]. However, Lehtiniemi et al. [153] identified size as the key factor influencing the ingestion of MPs by small predators. WWTP sludge byproducts used as fertilisers on agricultural land contain MFs that can persist in sludge and soil for up to 5 years [132], making them enduring terrestrial contaminants.

2.5.2 Health risk

MFs can potentially carry hazardous chemicals with negative impacts; however, their direct impact on humans is still not fully understood, although there is evidence that they reach the food web and humans. MPFs are the most prevalent MPs identified in air, terrestrial, and aquatic environments [28, 142, 154].

Ingestion of MPs by aquatic organisms has been documented across the marine food web, including turtles, seabirds, fish, crustaceans, and worms [155]. Laboratory studies have also reinforced that many other organisms, including zooplankton, can ingest MPs [45, 152]. Most related studies have documented MPs in the gut, an organ not directly consumed by humans. However, the risk of ingesting MPs contained within other tissues depends on the degree of uptake of MPs and their translocation, redistribution, and retention within other body tissues [156].

Moreover, the large surface area of MPs indicates that environmental pollutants may be absorbed onto the surface of particles, with the potential to be transferred into body tissues once ingested [155, 157]. These contaminated residues can absorb and concentrate organic pollutants ingested by marine fauna, which can be subsequently transferred to the food chain and potentially reach humans [43, 49, 52, 64].

The release of MFs from dried clothes could represent a source of airborne MPs. Increasing evidence of MFs in the atmosphere has recently been reported [53, 158]. Flying MFs inhaled by humans are deposited in the lung tissue and may lead to tumours. Several studies have shown that regular and prolonged exposures can cause respiratory inflammation, pulmonary fibrosis, and cancer [31]. Evidence of inhalation in a few studies has also found that human lung tissues contain MPs/MFs [66, 159]. There is also growing evidence of MPs/MFs being identified in different parts of humans and animals, such as the placenta [65], stool [160, 161], liver [162], and blood [163]. However, the direct impact of these pollutants on human health remains unclear [164]. Consequently, it is alarming where they are found reaching our bodies and the potential impact they can have on the chemical substances they carry.

3. Challenges

3.1 Undivided stakeholder responsibilities

The lack of a precise definition of MF pollution and the distinction between a problem and a challenge complicates the issue, potentially causing confusion and impeding effective communication and action among stakeholders [165]. This ambiguity hinders the setting of reduction targets, the creation of regulatory frameworks, and public awareness, leading to misunderstandings about the nature and extent of the problem. When stakeholder responsibilities are not clearly defined, solution development is hindered. When a problem is considered as everyone's responsibility, it often results in no single entity being able to address it independently. Consequently, there is no obvious starting point or clear leadership role to intervene and push into progress.

The complete elimination of MPFs/MFs is impractical because clothing wear and washing are essential human activities. Implementing global infrastructure to treat laundry or industrial wastewater and eliminate MF release is equally unfeasible. Consumer engagement is crucial; retailers must guide consumers on proper garment care and recommend washing machine filters to reduce microplastic release. Consumers can also reduce plastic use and choose sustainable products.

Industries and government initiatives play a critical role in mitigating MP contamination. Businesses across various sectors, including textile, washing and drying machine manufacturers, detergent producers, wastewater treatment technology providers, and machine builders, can reduce their MP footprint by investing in recycling technologies, adopting sustainable packaging, and implementing plastic waste management best practices.

3.2 MPs/MPFs/MFs as pollutants

The ubiquity of MPFs in the environment complicates accurate pollution assessment because MPs are found in nearly all environmental compartments, from remote mountain lakes to deep ocean trenches [1, 69]. This widespread distribution complicates the establishment of baseline pollution levels and tracking changes over time. Moreover, the potential for cross-contamination during sampling and analysis requires extremely rigorous protocols that can be time-consuming and resource-intensive. Moreover, the extensive distribution of MPs/MPFs in freshwater ecosystems, oceans, and seas, encompassing both water columns and sediments, further complicates the assessment of their spatial distribution and biotoxicity [166].

The diverse environments of MPFs pose challenges for researchers studying their ecosystem impact. Aquatic settings may affect microfibre behaviour, degradation, and toxicity in ways that are not fully understood, thus complicating comprehensive risk assessments and predictions of the long-term effects of MPF pollution on aquatic life and food webs. Increasing research has argued that current MPF reporting is likely underestimated because of methodologies and incomplete coverage of locations and pathways [154, 167–169].

MPF is of tiny size and not easily visible to the human eye. However, it is difficult to capture and handle. Therefore, specific high-end precision equipment such as micro-FTIR or micro-Raman is required [170] for identification and quantification, which is rarely available in a standard or commercial laboratory. Moreover, this has substantially impeded the starting point of commercial research.

3.3 Knowledge gap and mixed communications

Despite MFs being the main contributors to MP contamination, most studies focus on MPs [136]. Thousands of studies on MPs exist, but MFs have received limited attention. The MF information is often associated with MP pollution. For instance, Boucher and Friot [29] analysed MPs in over 2000 aquatic samples from freshwater and marine sources, facing challenges due to insufficient and undifferentiated MF data, which limited insights. This gap underscores the need for more MF research, accurate MF quantity reporting in ecosystems, and an understanding of material and factor influences. Existing strategies only address a limited aspect of the issue and should aim to reduce MF emissions throughout the product lifecycle. This data deficiency has resulted in industrial progress.

Most MF studies have targeted synthetic textile shedding during home laundering, overlooking MF release during manufacturing. Recent studies have shown that removing MFs during fabric production reduces garment shedding risk [37, 74, 171]. Modifying construction parameters to address shedding earlier in the lifecycle is more effective than post-shedding removal when fibres are environmentally dispersed; thus, industry stakeholders must understand and follow best practices in engineering textile materials [104, 172].

Moreover, a significant amount of current research focuses on textile materials or clothing that are randomly available in the marketplace [59, 102, 108, 170, 173]. These studies often overlook key textile parameters, such as the number of filaments and textile finishes, which can significantly affect their performance. Insufficient information on manufacturing processes and products limits our understanding of lifecycle impacts and hotspots [22]. This knowledge gap encompasses the entire lifecycle of textiles and other products that contribute to MF pollution. It is challenging to implement targeted interventions without understanding how various manufacturing processes, materials, and product designs contribute to microfibre shedding. This lack of information also impedes the development of sustainable production methods and materials that could reduce MF release, rendering industrial solutions questionable.

Pirc et al. [75] found that after five washes, microfibre (MF) release stabilised, regardless of washing conditions, with a significant release during the first wash, regardless of detergent, detergent, or detergent with a softener. With the % weight loss method, the release dropped by 64–74% after the initial wash. Conversely, Almroth et al. [101] counted the number of fibres released in subsequent washes, which is considered more reliable [111]. Further washing cycles showed no clear correlation with the primary fabric characteristics, likely owing to inadequate textile parameters for accurate assessment. Limited research has investigated these aspects and the contamination potential of initially washed fabrics. Although pre-washing garments have been proposed [174], the industry may resist this because of the need for additional processing and infrastructure.

The Changing Markets Foundation [175] argued that the industry is shifting its plastic pollution narrative by claiming that all MFs, regardless of origin, are equally problematic, which contradicts scientific evidence highlighting the specific dangers of MPs. The widespread presence of natural MFs in various environments raises concerns, indicating that MP pollution threats extend beyond synthetic fibres. The industry faces conflicting messages from NGOs, academia, and industrial bodies, such as TMC, complicating efforts to address MP pollution and broader fibre fragmentation issues, thereby delaying remedial actions.

3.4 Lack of standardised test methodology

A significant challenge is the absence of standardised test methods and reporting units to detect and quantify MPFs, particularly in industrial wastewater effluents [167]. This inconsistency obstructs accurate comparisons and formulation of practical solutions. The lack of uniform protocols hampers researchers and policymakers in evaluating the actual scope of MPF pollution and comparing the findings across various studies and regions. This standardisation deficit also complicates the establishment of regulatory limits and benchmarks for MF emissions.

The development of novel testing techniques requires industry cooperation, accelerated by stakeholder demands and regulatory actions for harmonised measures. Current standardised methods do not consider other particulate contaminants, fibre-type distinctions, or length distributions. The precision and reliability of the control parameters need further refinement, depending on the polymer attributes absent in natural fibres. Some protocols use solvents to remove contaminants, including natural fibres, which can dissolve or eliminate organic matter. Therefore, the methodology formulation must include fibre types prevalent in the textile industry.

Developing methodologies for measuring MF release is crucial for evaluating solutions and making science-based decisions. Standardised methods can enhance regulatory accountability and set maximum thresholds for environmental MF release, thus effectively mitigating this issue. A more unified approach using AATCC TM212 [119], ISO/DIS 4484–1 [120, 176], and the TMC method-2019 [177] is now available for the industry to evaluate MFs released from laundry. There is continued demand to expand its scope to cover drying, air, wastewater, and sludge. Techniques capable of separately enumerating and quantifying the size of blended synthetic and natural fibres are crucial, given their inherently distinct impact and toxicity mechanisms.

3.5 Lack of legislation

Despite growing concerns about MPF pollution, specific legislation remains scarce. Experts recommend incorporating MPF environmental policies into textile regulations to address the issue at its source [178]. The current legislation on MPF pollution is nascent, with most regulations focusing on MPs rather than MPFs. For instance, although several countries have banned microbeads, these policies do not address other MPs, including fibres [179].

A few countries have made limited legislative efforts on MPF. In 2018, California Assembly Bill No. 2379 and Connecticut General Assembly No. 341 did not advance beyond the proposals. France, however, took a pioneering step in 2020 with Law No. 2020–105, known as the Anti-Waste for a Circular Economy (AGEC) Law. This law, which includes Article 79 targeting MPF pollution from textiles, aims to reduce waste and promote a circular economy through Extended Producer Responsibility (EPR) and eco-design requirements. Since January 2025, new washing machines in France must include MPF filters. In September 2023, the European Parliament revisited a proposal to mandate the use of MP filters in washing machines at the EU level, citing environmental and health risks. They urged the Commission to consider EU-wide regulations, with ongoing discussions about feasibility and implementation timelines across member states [180].

The UN Alliance for Sustainable Fashion, launched in 2019, promotes Sustainable Development Goals (SDGs) within the fashion industry, encouraging stakeholder alignment. Thus, reducing emissions in the fashion value chain remains a significant challenge. The Cross-Industry Agreement seeks to prevent MP release into aquatic environments during synthetic textile washing, with the European textile industry developing a standardised testing method. TMC establishes standards for post-laundry microfibre release testing and requires members to adhere to the minimum quality standards for a data portal. TMC and the ZDH Foundation (ZDHC) have also developed voluntary guidelines to control MF) release from wastewater. Despite these voluntary initiatives, progress remains modest, and non-governmental organisations (NGOs) [175] often question the transparency and methodologies used in these efforts.

4. Priorities

4.1 Stakeholder collaboration and industry-led initiative

Of all priorities, stakeholder collaboration through an Industry-led Initiative should be the most important and have an overarching role in leading or guiding action plans. There used to be Industry Action for Microfibres at an early stage to address this problem with multiple stakeholders, including associated sectors, and they published an assessment report [181]. However, this group did not move forward. The Microfibre Consortium (TMC), set up in 2021 with its 2030 roadmap, has become the only global industry-led and science-led initiative focusing on addressing MF pollution by creating change through design, development, and manufacturing. Compared to other sustainability initiatives, TMC is the youngest. They published a recent progress report on their achievements in upgrading the testing data portal for the TMC TM212 method, hitting over 850 fabric performances on MF release after the first launch.

Table 4 summarises the selected industry sustainability initiatives related to the MF agenda and are commonly referenced in the brand and NGO reports. However, each has its own strategy and delivery method. Canopystyle has been rated as the most successful organisation, as it has the second highest number of partners' support and has received several awards for its impactful work in sustainable fashion and forest conservation. Notably, it won the BF + DA positive impact award for leadership to create a better supply chain in November 2016. Additionally, Canopy, the organisation behind Canopystyle, was recognised with the US$3 million Climate Breakthrough Award in 2021. It is just about 11 years, but in terms of partners, it reached 500, which is significantly more than Cascale, which was established nearly 15 years ago, although they are also recognised with the Eco-Steward of the Year in 2021 (when they were formerly known as SAC). This recognition highlights their efforts to drive sustainability and transparency across the apparel and footwear industries. Similar to TMC, they have a small team that focuses on shifting the retailer-sourcing policy to embrace a public commitment to resolve the solution. At the same time, the progress of fibre producers will be assessed using their annual hot-button reports. Their strategic model is unique but effective compared with their sphere of impact.

Furthermore, the majority of these organisations possess a governance framework that includes a Board of Directors comprising diverse stakeholders from their communities. This structure ensures that the interests of various groups are well represented while maintaining a focus on swift progress towards established objectives. The inclusion of external representatives in guiding the team instils greater confidence within the industry to participate, as these members can influence and shape the direction of the organisation's activities.

Industry Initiatives	**The Microfibre Consortium (TMC)**	**UN Fashion Industry Charter for Climate Action (UNFCCC)**	**ZDHC Foundation (ZDHC)**	**CanopyStyle by Canopy**	**Cascale (formerly SAC)**	**The Ellen MacArthur Foundation (EMF)**	**Textile Exchange (TE)**
Establish	2021 (3 yrs)	2018 (6 years)	2015 (9 yrs)	2013 (11 yrs)	2009 (15 yrs)	2010 (14 yrs)	2002 (22 yrs)
Purpose	A non-profit, science-led organisation with a vision to work towards zero impact from fibre fragmentation from textiles to the natural environment.	An industry-led body of signatories working collaboratively to drive climate action in the sector and quantify, track, and publicly report GHG emissions.	A multi-stakeholder organisation with a program leading the fashion industry to eliminate harmful chemicals from its global supply chain to protect workers, consumers and the planet's ecosystems.	A collaboration between Canopy, fashion brands, retailers, designers, and their MMCF suppliers to follow the product thread and keep Ancient and Endangered Forests out of textiles.	A global, non-profit alliance working towards a shared vision of an industry that gives back more than it takes to the planet and its people.	A charity committed to creating a circular economy designed to eliminate waste and pollution, circulate products and materials (at their highest value), and regenerate nature. It's an economic system that delivers better outcomes for people and the environment.	A global non-profit driving beneficial impacts on climate and nature across the fashion, textile, and apparel industries.
Stake-holders	Brands, retailers, manufacturers, research institutions, multi-stakeholder initiatives, Industry associations, and fabric testing laboratories.	Any company (e g brand, retailer, supplier) that is Professionally engaged in the fashion sector.	Brands, Suppliers, Solution Providers and Chemical Suppliers.	Fashion brands, retailers, designers, and MMCF suppliers.	Brands, retailers, manufacturers, sourcing agents, service providers, trade associations, NGOs, and academic institutions.	Business, government, academia, designers, innovators, NGOs, and others.	Brands, manufacturers and farmers
Members	101 signatories	~100 signatories	320 signatories	>500 fashion industry partners	303 members	N.A. (Thousands of network stakeholders)	>800 members
Business Model	Signatory Fee + donations	No Fee (supported by UN)	Signatory Fee + program operational income + donations	No corporate fee + donations	Membership fee + program operational income + donations	Corporate Membership + donations	Membership Fee + program operational income + donations
Governance	Not quoted	UN framework	Multistakeholder Board	External Board	Multistakeholder Board	Not quoted	Multistakeholder Board

Industry Initiatives	The Microfibre Consortium (TMC)	UN Fashion Industry Charter for Climate Action (UNFCCC)	ZDHC Foundation (ZDHC)	CanopyStyle by Canopy	Cascale (formerly SAC)	The Ellen MacArthur Foundation (EMF)	Textile Exchange (TE)
Related Impact (quote from their latest report)	Created the TMC test method and the Microfibre Data Portal, sharing test results with more than 800 fabrics for signatories.	80% have SBTi-approved targets, and over 50% are validated. 32% aggregated reduction for scope 3 from 2023 to 2030.	98% of suppliers tested wastewater in 2020 had no detections of restricted substances from the ZDHC MRSL parameters.	29% low-carbon, circular Next Gen lines, 71% of the world's MMCF producers (or 54% of global production) have achieved green shirt.	24,000 HIGG users, 59.7% of corporate members, committed to SBTs.	1.8 billion tons of potential savings by 2030 (3x global production of a single piece of plastic, clothing and consumer electronics.	Electronic transactions went live as eTrackit for certified materials on the platform as the foundation for validating the chain of custody with full transparency.
Strategies	We take a holistic approach to creating change through design & development and in manufacturing.	Through collaborative efforts, information sharing, partnerships, shared principles and goals to accelerate decarbonization efforts.	Three focus areas are interlinked with platforms and solutions to accelerate the implementations.	Transform outdated supply chains, reduce company risk, catalyse Next Generation Solutions, and engage in collective advocacy that advances forest conservation.	Create a global approach for measuring and evaluating the social and environmental impacts of value chains and products, improve sustainability performance and engage with value chain partners to scale systemic change.	Change Business Practices, Shift Mindsets, Create Enabling Conditions, common roadmap	Inspires and equips people to accelerate the adoption of preferred materials through clear and actionable guidance and provides tools, resources, and initiatives to drive change.
Sources	https://www.microfibreconsortium.com/	https://unfccc.int/climate-action/sectoral-engagement-for-climate-action/fashion-charter	https://www.roadmaptozero.com/	https://canopyplanet.org/change-together/canopystyle	https://cascale.org/	https://www.ellenmacarthurfoundation.org/	https://textileexchange.org/about/

Table 4.
Selected Industry Sustainability Initiatives which are related to the microfibre agenda and are referenced in the brand and NGOs reports.

Owing to the vast variations in textile materials, there is instinctively considerable variation in the sample specimens used, which makes the comparison inconclusive. Researchers from a non-textile background may collaborate with industry or academic disciplines to support the selection of fabrications that have an enormous impact and can provide detailed textile parameters to be selected. Therefore, this initiative needs to work closely with academia. Cascale has a material sustainability index (MSI) tool, which should be an excellent fit for collecting textile parameters and working together to measure MF release without duplicating some groundwork.

Indeed, MPF/MF pollution is a diverse cross-cutting issue for the entire textile and clothing industry and its associated supporting industries, including wastewater treatment, fibre and textile, garment or product manufacturing, washing machines, detergents, and testing services. It is relevant to the fibre and material manufacturing sector when addressing the impact of fibre type and textile manufacturing. Conversely, stakeholders and processes influenced by product design become the most relevant when addressing sewing processes. To enhance the effectiveness of WWTPs, collaboration with technology and machinery providers is necessary. The detergent and washing machine manufacturing industries should be integrated into the solution. To resolve this issue comprehensively, it is essential to consider all possibilities; therefore, it is crucial to drive improvements through all potential options.

The ability to connect all stakeholders determines the speed at which solutions can be scaled and implemented. Connecting diverse stakeholders is crucial for rapid scaling and implementation of solutions across various sectors. This interconnectedness facilitates the exchange of ideas, resources, and expertise, thus enabling a more efficient and effective problem-solving process. Furthermore, strong connections between stakeholders foster collaboration and synergy, leading to innovative solutions that may not be possible through isolated efforts. This network of relationships can also help streamline decision-making processes, reduce redundancies, and optimise resource allocation. Consequently, the speed at which solutions can be scaled and implemented is significantly enhanced, allowing for more timely responses to this type of sustainability issue.

By bringing together individuals and organisations from different backgrounds, such as government agencies, private businesses, non-profit organisations, and academic institutions, a comprehensive approach to addressing complex challenges can be developed. Governments, industries, communities, and research institutions must establish partnerships and collaborate to identify solutions for MPF/MF pollution. By combining resources, expertise, and knowledge to pursue a common objective, novel solutions can be generated, and systemic changes can be affected. International cooperation is particularly crucial for addressing transboundary issues, such as MP/MPF pollution, which transcends national boundaries and necessitates coordinated action on a global scale. The initiatives are still driven by the Global North, irrespective of whether manufacturing is in the Global South. Geographical collaboration has been regionally limited.

While allocating responsibilities to complex issues is complicated, involving various stakeholder groups based on their interests and roles in crafting workable solutions is crucial. When framed as MP/MPF pollution, it is probable that supply chain and activewear retailers who depend on synthetic materials as their primary resources bear significant responsibility and face the highest business stress and risk. Their mitigation approaches will likely differ from those relying on natural fibres, such as cotton, as their concerns revolve more around altering chemical properties, which makes it less easy to decompose and potentially harmful substances that may be released into the environment; consequently, the solution primarily involves addressing the chemicals used in manufacturing. The resolution of MPF pollution is unlikely to occur rapidly unless synthetic fibres are substituted, owing to their cumulative nature, non-biodegradability, and continuous dispersal, which render their removal from the environment impractical. Consequently, the level of stakeholder engagement can vary considerably. Framing the issue as a widespread microfibre pollution problem stemming from fibre composition may complicate the process of involving relevant stakeholder groups without yielding substantial collaborative benefits. It is essential to strike a balance among these strategic perspectives.

4.2 Raising awareness

Awareness and education among all MF release supply chain stakeholders are crucial for developing suitable mitigation measures [22, 175, 182]. Every stakeholder group acknowledges its possible contribution in resolving this issue. Industry measures may involve altering the material choices for design and production. Consumers can express their concerns to retailers and manufacturers and reduce fibre loss at home by decreasing the effluent from domestic washing machines.

MFs are minuscule and vary in density, leading to their spread across different layers of the water column [155]. Consequently, the complete removal of MFs from oceans is impractical. Currently, the accumulation of MPs in the environment surpasses their removal rate [183]. Although some studies have identified bacterial species capable of degrading plastic polymers such as PET in soil [184], these microbes are not widely used for cleanup because of the uncertain risks of releasing bacteria into the environment. Recently, a solution for cleaning ocean plastic debris has been proposed [185]; however, it is more effective for MFs longer than 1000 μm, whereas most MFs are shorter, highlighting a significant gap in cleanup efforts. Thus, stakeholders should adopt a precautionary approach to reduce MF generation and prevent its entry into aquatic environments.

Consumer-oriented solutions encouraging behavioural change include less frequent washes or increased use of laundry devices, such as coral balls and grubby bags, to capture or reduce filter releases through the washing machine to the effluent. Filters have been proven to be effective in capturing MFs by 30–60% from the washing machine outlet [91, 113]. This solution requires commitment and motivation. Although raising consumer awareness of laundering behaviours can only offer a limited short-term resolution [181], legislation is always the most potent tool for transformational and lasting change.

4.3 Sustainable manufacturing technology

Product design improvement is advised by considering critical parameters that significantly affect the MFs released during washing [186]. **Table 2** shows that longer fibres, higher yarn twists, coarser yarn counts, increased fabric density, PA coatings, and fewer textile auxiliaries reduced the MF release and enhanced the shedding performance. These steps are vital during the design phase. Decision-makers require further support to encourage designers and manufacturers to use improved parameters with reduced MF release. Unbiased testing is essential to measure the effectiveness of these improvements.

Recommendations for manufacturing processes, such as machinery type, processing time, and temperature, remain unexplored. Mitigating MF shedding may involve enhancing the cohesive bonds between the fibres produced. However, there is a lack of conclusive data on its effectiveness. Some suggestions include pre-washing, air filtration, and exhaustion at the site [101]. However, this approach should be cautiously implemented, as industrial facilities typically exercise better control over releases than household laundries. Proper management and disposal methods are essential to prevent environmental contamination.

Advanced technologies such as Pangaia have partnered with MTIX's multiplexed laser surface enhancement (MLSE®) technology to modify the surfaces of fibres within a fabric in a way that prevents microfiber shedding [187]. MLSE® rapidly alters textile surfaces in a high-energy reaction zone generated from the combined high-power pulsed ultraviolet laser and high-frequency electrical discharge plasma [188]. The nano-level treatment reconstituted would-be microfibres back into the fibres, resulting in stronger, cleaner fibres with restored integrity and uniformity.

Other solutions have focused on innovation of renewable sources, including alternative bio-based and biodegradable materials. Mango Materials currently produces poly-3-hydroxybutyrate (P3HB), a member of the PHA family. It is a biopolyester with properties similar to polypropylene [189]. Werewool uses biotechnology to design biodegradable fibres at the DNA level with inherent properties, such as colour, moisture management, and stretch, that meet performance requirements. Similarly, Tandem Repeat Technologies [190] use genetic sequencing and synthetic biology to produce a new fibre based on a unique protein structure initially found in the tentacles of squid, which is 100% biodegradable and recyclable. Another breakthrough is the use of plants with an inexhaustible abundance on Earth. By leveraging their strength, Natural Fibre Welding [191] can provide a sustainable and circular material that uses this renewable resource. However, these solutions are still at an early stage, entering the niche market at premium prices that are yet to scale. Continuing development in these areas should be encouraged by supportive adoption pathways to become widely available.

On the other hand, transforming plastic bottles into textiles for clothing is inadvisable as it disrupts the underperforming bottle-to-bottle recycling process. Clothing from these bottles cannot be recycled into high-quality materials due to limitations in textile recycling methods, making landfills or incineration more likely. This method does not effectively tackle MP pollution or waste management issues [175]. The industry should prioritise advanced technologies that facilitate large-scale textile-to-textile recycling to reduce environmental waste.

Yousef et al. [192] piloted a method for converting dryer-produced Lint-Microfibres (LMFs) into energy products via thermal processing. This innovative approach transforms MF waste into a valuable energy resource, potentially lowering industrial carbon emissions. As a result, LMFs can be seen as a renewable energy source, advancing the textile industry towards a circular economy.

4.4 Wastewater treatment technologies advancement

Water stress and MF pollution can be substantially mitigated using membrane bioreactors (MBR) and Zero Liquid Discharge (ZLD) systems, with reports indicating a reduction in MP discharge of almost 97% [123]. While MBR technology is potentially the most efficacious method for removing MF from wastewater, the selection of an appropriate technique is contingent upon the characteristics of the MPs/MPFs, including their material, morphology, size, and density, highlighting the need for additional research [193]. Zhou et al. [194] proposed dissolved air flotation (DAF) as a cost-effective alternative to reverse osmosis and ZLD for MF separation, given the lower density of MFs compared with effluent. However, the substantial investment and energy requirements of advanced technologies may contribute to climate change and restrict their applications in industrial WWTPs. It should be noted that membrane technology is not explicitly designed for MF removal [195]. Consequently, there is a pressing need for more economical and efficient methods to prevent MFs from entering the environment.

Advanced WWTP technologies can effectively prevent MF discharge as effluents, but their importance remains essential for tackling MF pollution unless dry production processes can completely replace conventional wet methods. Increasing the amount of recycled water within WWTPs and reducing wastewater discharge can be immediate solutions. Innovations in dry finishing technologies, such as using carbon dioxide as a dyeing carrier, are cleaner than wet processes and can simultaneously address MFs and water pollution, potentially accelerating investment decisions. However, barriers such as a lack of knowledge and test methodology and the inability to include MF release requirements in sourcing policies may have limited financial incentives and delayed industrial transformation. Beyond standardising test methodologies, incentives such as green funds supporting new technologies are crucial for reducing MF discharge through this aquatic pathway.

The transformation of the textile industry to address MPF/MF pollution is hindered by limited knowledge, testing methods, integration of MF criteria in retail sourcing, and financial incentives. The adoption of sustainable technologies and green production can promote quicker investment decisions. However, advanced technology has limitations unless MF pollution is addressed. Ozone dyeing, a waterless process, supports cleaner production but may lead to higher MF concentrations in sludge, thus affecting terrestrial environments. As WWTP retention rates improve, terrestrial and airborne MF levels MFs could surpass aquatic emissions [196]. Effective sludge management is crucial for preventing the transfer of this issue to soil ecosystems. Incinerating textile sludge at 800°C removes heavy metals, reduces volume, and decomposes all fibres. The resulting ash can be used in construction blocks to prevent MF transfer to terrestrial environments and to transform MFs into valuable fuels [197]. Therefore, careful sludge management supports systematic transformational changes without transferring the issue from one ecosystem to another.

4.5 Standardise measurement

Despite the few commercial test methods outlined in **Table 3**, there is a lack of suitable techniques for assessing MF release from yarns, garments, or WWTPs. The absence of comparable measurements hinders effective management and control, emphasising the need for a standardised testing methodology. Addressing this knowledge gap requires collaboration beyond the academic and industrial spheres. Therefore, it is crucial to develop scalable, diverse, and innovative solutions for practical applications. The Microfibre Consortium (TMC) exemplifies the efforts to unite industry expertise, academia, and stakeholders in developing solutions. To achieve a scalable impact, the results need to be shared for the entire industry to implement, rather than being limited to signatories. The development of necessary infrastructure and tools may necessitate additional financial support from governmental bodies, environmental funds, or charitable organisations, supplementing industry contributions.

Understanding MF release is essential for combating textile pollution and developing precise mitigation strategies throughout the lifecycle of a product. Despite methodological challenges, ongoing research has enhanced the quantification and assessment of the efficacy of interventions. Accurate measurements enable the evaluation of various textiles and mitigation approaches, thereby identifying critical stages in the textile lifecycle for emission reduction [198]. However, methodological disparities have led to varied findings, and the impact of textile variables on MF release remains complex, warranting further investigation.

Inconsistent presentation of results, particularly regarding measurement units, complicates data analysis and impedes definitive conclusions. This complexity forces industries and governments to make decisions based on limited information, underscoring the urgent need for standardised methodologies to obtain accurate data and develop informed strategies. The absence of uniform testing approaches in the textile industry necessitates global collaboration to establish common standards and to provide reliable data for manufacturing improvements. Various reporting standards and insufficient method transparency hinder research progress. MF studies should also adopt standardised methods to enhance reproducibility, similar to efforts towards unified reporting guidelines in MP research [199]. This would improve the reliability of impact assessments and formulation of mitigation strategies, thereby supporting regulatory initiatives.

4.6 Responsible consumption and circularity

Behavioural changes are recommended at the consumer level. These include reducing consumption and extending garment life which can represent part of the solution. Reduced consumption and manufacturing would reduce the quantity of MFs released into the environment by using fewer materials. However, a significant challenge remains involving simultaneous and systematic changes across consumers, multiple stakeholders, and government efforts. According to the Ellen MacArthur Foundation [24], less than 1% of clothing is recycled into apparel. To achieve circularity, it is essential to establish an infrastructure that enables consumers to recycle their clothing and textile items easily. Transforming consumer behaviour requires concerted efforts from retailers and the implementation of local incentive programmes.

The laundry parameters outlined in **Table 1** unequivocally determined that laundry settings play a crucial role in minimising MF release. This requires consumer education, behavioural changes, and the functions and built-in features of washing machines. Consumers are advised to wash in cold water with fewer cycles and launder clothes when necessary and at full load [84, 89], which also indirectly extends a garment's lifetime irrespective of conflicting parameters when detergents and softeners are used. This requires changing consumer behaviour when purchasing and caring for clothes. Shifting to gentler wash and dry conditions can be the most economically advantageous option, as it simply entails adjusting the current machine settings and reducing energy usage.

Options that incur additional costs to replace a front-loading washing machine that generates fewer MFs or purchases but high-quality clothing may be more challenging for most consumers to adapt; this type of machine is less common in certain countries because of its size. Manufacturers of washing machines, tumble dryers, detergents, and softeners should also take responsibility for and evaluate their contribution to reducing the MFs released by improving the design of their products and communicating better options and laundry conditions to their customers.

4.7 Legislations and voluntary intervention

The EU Strategy for Textiles and UN Alliance for Sustainable Fashion support efforts to address MPF pollution. France is the only EU country with an existing Extended Producer Responsibility (EPR) scheme for textiles. The Netherlands advocated an EU-wide EPR for textiles and introduced its own scheme in 2023. Sweden enacted legislation for EPR for textiles starting 1 January 2022, with full implementation by 2024 [200]. These measures aim to reduce reliance on virgin synthetic materials and eliminate MPF release using recycled or alternative sources. The EU framework could transfer costs to textile producers and the fashion industry, possibly by taxing the most polluting fabrics and garments, thereby promoting materials with minimal MF release.

Research indicates that effective wastewater treatment, such as zero liquid discharge (ZLD) and tertiary methods, require infrastructure investment. Filtering devices on washing machines and dryers can also help reduce the burden on wastewater treatment systems. For air quality monitoring, microfibres (MFs) should be classified as fine particulate matter. Monitoring is essential to reduce uncertainty about their sources and fate, although methodologies and standardisation of MPF/MF measurements remain challenging, delaying the implementation of regulations. The voluntary industry best practices should be encouraged as an alternative to legislation to address MPF/MF pollution.

Current textile legislation lacks environmental provisions or specific regulations on MPF/MF pollution [178]. Environmental policies on microplastics (MP) should be integrated into textile regulations. A multidisciplinary approach combining research, regulations, and public engagement is crucial for protecting marine life and human health from microplastic contamination [201]. Future policies should address the entire lifecycle of plastics, promote a circular economy, and implement strategies to limit the environmental plastic input [202]. Strengthening laws and regulations specific to MPF pollution is also recommended [163].

5. Conclusion

Scientific findings suggest that MPs/MPFs present the highest environmental and human health risks among MFs, thereby guiding necessary prioritisation decisions. The replacement of MPFs with natural or bio-based alternatives is not required to avoid regrettable solutions. Emphasising reduced consumption and demand for MPFs and promoting a circular model is a more viable approach to prevent problem transfer. Addressing these challenges requires a holistic approach involving multiple stakeholders, including industry awareness, academic investigations, and targeted government interventions [22]. Collaboration across sectors is crucial for developing comprehensive solutions that address the entire lifecycle of the products that contribute to MF pollution. Industry involvement is essential for implementing manufacturing and product design changes to reduce MF shedding at the source.

Consistent protocols for measuring microfibres in wastewater treatment plants can help to define the maximum limits for MF release and support legislation [167]. Standardised methods for quantifying MF in wastewater effluents would provide a solid foundation for regulatory efforts and enable accurate assessments of the effectiveness of treatment technologies, thereby facilitating targeted and effective pollution control strategies.

Exploring source-directed interventions in textile design and manufacturing, such as adjusting fabric structure and cutting techniques, can reduce MF release [203]. This includes developing new textile materials that shed fewer fibres, optimising fabric construction to minimise fibre loss, and improving manufacturing processes to reduce waste and microfibre generation. These approaches can significantly reduce MF emissions and complement end-of-pipe solutions such as improved wastewater treatment. Voluntary approaches with industrial guidelines on production best practices provided by industry-led and science-led information are essential and more effective than regulatory intervention, as it is not feasible for the granularity of specifications that manufacturers can apply practically.

Tackling MF pollution requires a multifaceted approach addressing standardisation, accurate detection, and quantification while focusing on prevention through improved textile design and manufacturing processes. Collaborative efforts among stakeholders are crucial for implementing systematic changes and developing effective strategies to mitigate the environmental impact of microfibre pollution. This comprehensive approach should encompass research, policy development, industry innovation, and public education to create a more sustainable textile industry and to protect aquatic ecosystems from harmful MF pollution. Governments, industries, communities, and research institutions must establish partnerships and collaborate to identify solutions to MP pollution. By combining resources, expertise, and knowledge to pursue a common objective, novel solutions can be generated, and systemic changes can be affected. International cooperation is crucial for addressing transboundary MF pollution, which transcends national boundaries and necessitates coordinated action on a global scale.

Author details

Carmen Ka Man Chan[1*], James Kar Hei Fang[2], Fei Bin[1] and Chi Wai Kan[1]

1 School of Fashion and Textiles, The Hong Kong Polytechnic University, Kowloon, Hong Kong

2 Faculty of Science, The Hong Kong Polytechnic University, Kowloon, Hong Kong

*Address all correspondence to: carmenkm-itc.chan@connect.polyu.hk

References

[1] Napper IE, Davies BFR, Clifford H, Elvin S, Koldewey HJ, Mayewski PA, et al. Reaching new heights in plastic pollution: Preliminary findings of microplastics on Mount Everest. One Earth. 2020b;**3**(5):621-630. DOI: 10.1016/j.oneear.2020.10.020

[2] Rebelein A, Int-Veen I, Kammann U, Scharsack JP. Microplastic fibers — Underestimated threat to aquatic organisms? Science of the Total Environment. 2021;**777**:146045. DOI: 10.1016/j.scitotenv.2021.146045

[3] Jain R, Gaur A, Suravajhala R, Chauhan U, Pant M, Tripathi V, et al. Microplastic pollution: Understanding microbial degradation and strategies for pollutant reduction. Science of the Total Environment [Internet]. 2023;**905**:167098. DOI: 10.1016/j.scitotenv.2023.167098

[4] Manosuthi J. Microplastic pollution: The real potential threats to human being. Science Insights. 2024;**44**(2):1263-1272. DOI: 10.15354/si.24.re925

[5] Arthur C, Baker J, Bamford H. In: National Oceanic and Atmospheric Administration Technical Memorandum NOS-OR&R-30, editor. Proceedings of the International Research Workshop on the Occurrence, Effects, and Fate of Microplastic Marine Debris. Washington, USA: NOAA Marine Debris Division; 2008. p. 530. Issue January. Available from: https://marinedebris.noaa.gov/proceedings-international-research-workshop-microplastic-marine-debris

[6] Costa MF, Do Ivar Sul JA, Silva-Cavalcanti JS, Araújo MCB, Spengler Â, Tourinho PS. On the importance of size of plastic fragments and pellets on the strandline: A snapshot of a Brazilian beach. Environmental Monitoring and Assessment. 2010;**168**(1-4):299-304. DOI: 10.1007/s10661-009-1113-4

[7] Thompson RC, Olson Y, Mitchell RP, Davis A, Rowland SJ, John AWG, et al. Lost at sea: Where is all the plastic? Science. 2004;**304**(5672):838. DOI: 10.1126/science.1094559

[8] GESAMP. Sources, Fate and Effects of Microplastics in the Marine Environment (Part 1)|GESAMP [Internet]. GESAMP; 2015. Available from: http://www.gesamp.org/publications/reports-and-studies-no-90

[9] Kabir AHME, Sekine M, Imai T, Yamamoto K, Kanno A, Higuchi T. Assessing small-scale freshwater microplastics pollution, land-use, source-to-sink conduits, and pollution risks: Perspectives from Japanese rivers polluted with microplastics. Science of the Total Environment. 2021;**768**:144655. DOI: 10.1016/j.scitotenv.2020.144655

[10] Agbekpornu P, Kevudo I, Achparaki M, Thessalonikeos E, Tsoukali H, Mastrogianni O, et al. The Risks of Microplastic Pollution in the Aquatic Ecosystem. Vol. Vol. i (Issue tourism). London, UK: IntechOpen; 2012. p. 13. DOI: 10.5772/intechopen.108717

[11] Barrows APW, Neumann CA, Berger ML, Shaw SD. Grab: Vs. neuston tow net: A microplastic sampling performance comparison and possible advances in the field. Analytical Methods. 2017;**9**(9):1446-1453. DOI: 10.1039/c6ay02387h

[12] Fischer EK, Paglialonga L, Czech E, Tamminga M. Microplastic pollution

in lakes and Lake shoreline sediments: a case study on Lake Bolsena and Lake Chiusi (Central Italy) *. Environmental Pollution. 2016;**213**:648-657. DOI: 10.1016/j.envpol.2016.03.012

[13] Zhao S, Zhu L, Wang T, Li D. Suspended microplastics in the surface water of the Yangtze estuary system, China: First observations on occurrence, distribution. Marine Pollution Bulletin. 2014;**86**(1-2):562-568. DOI: 10.1016/j.marpolbul.2014.06.032

[14] Henry B, Laitala K, Klepp IG. Microplastic pollution from textiles: A literature review (issue 1). In: Consumption Research Norway - SIFO. Norway: Oslo and Akershus University College of Applied Sciences; 2018a

[15] Napper IE, Thompson RC. Release of synthetic microplastic fibres from domestic washing machines: Effects of fabric type and washing conditions. Marine Pollution Bulletin. 2016;**112**(1-2):39-45. DOI: 10.1016/j.marpolbul.2016.09.025

[16] Mishra S, Singh RP, Rath CC, Das AP. Synthetic microfibres: Source, transport, and remediation. Journal of Water Process Engineering. 2020;**38**(September):101612. DOI: 10.1016/j.jwpe.2020.101612

[17] Akanyange SN, Lyu X, Zhao X, Li X, Zhang Y, Crittenden JC, et al. Does microplastic really represent a threat? A review of the atmospheric contamination sources and potential impacts. Science of the Total Environment. 2021;**777**(10):51211-51233. DOI: 10.1016/j.scitotenv.2021.146020

[18] Liu J, Liu Q, An L, Wang M, Yang Q, Zhu B, et al. Microfiber pollution in the earth system. Reviews of Environmental Contamination and Toxicology. 2022;**260**(1). DOI: 10.1007/s44169-022-00015-9

[19] Xu YY, Chan FKS, Stanton T, Johnson MF, Kay P, He J, et al. Synthesis of dominant plastic microfibre prevalence and pollution control feasibility in Chinese freshwater environments. Science of the Total Environment. 2021a;**783**:146863. DOI: 10.1016/j.scitotenv.2021.146863

[20] Liang Y, Lehmann A, Yang G, Leifheit EF, Rillig MC. Effects of microplastic fibers on soil aggregation and enzyme activities are organic matter dependent. Frontiers in Environmental Science. 2021;**9**. DOI: 10.3389/fenvs.2021.650155

[21] Bank MS, Hansson SV. The microplastic cycle: An introduction to a complex issue. Microplastic in the Environment: Pattern and Process. 2021:1-16. DOI: 10.1007/978-3-030-78627-4_1

[22] Chan CK-MM, Lo CK-YY, Kan C-WW. A systematic literature review for addressing microplastic fibre pollution: Urgency and opportunities. Water (Switzerland). 2024;**16**(14):1-29. DOI: 10.3390/w16141988

[23] Textile Exchange. Materials Market Report 2023 (Issue December). USA: Textile Exchange; 2023. Available from: https://textileexchange.org/app/uploads/2023/11/Materials-Market-Report-2023.pdf

[24] EMAF. A New Textiles Economy: Redesigning Fashion's Future. Isle of Wight, United Kingdom: Ellen MacArthur Foundation; 2017. Available from: https://www.ellenmacarthurfoundation.org/publications/a-new-textiles

[25] Rathinamoorthy R, Balasaraswathi SR. Investigations on the interactive effect of laundry parameters on microfiber release from

polyester-knitted fabric. Fibers and Polymers. 2022;**23**(7):2052-2061. DOI: 10.1007/s12221-022-4929-y

[26] Huang Y, Tian M, Jin F, Chen M, Liu Z, He S, et al. Coupled effects of urbanization level and dam on microplastics in surface waters in a coastal watershed of Southeast China. Marine Pollution Bulletin. 2020;**154**. DOI: 10.1016/j.marpolbul.2020.111089

[27] Browne MA, Crump P, Niven SJ, Teuten E, Tonkin A, Galloway T, et al. Accumulation of microplastic on shorelines wolrdwide: Sources and sinks. Environmental Science and Technology. 2011;**45**(21):9175-9179. DOI: 10.1021/es201811s

[28] Barrows APW, Cathey SE, Petersen CW. Marine environment microfiber contamination: Global patterns and the diversity of microparticle origins. Environmental Pollution. 2018;**237**:275-284. DOI: 10.1016/j.envpol.2018.02.062

[29] Boucher J, Friot D. Primary microplastics in the oceans: A global evaluation of sources. Primary Microplastics in the Oceans: A Global Evaluation of Sources. 2017;**43**. DOI: 10.2305/iucn.ch.2017.01.en

[30] Woodall LC, Sanchez-Vidal A, Canals M, Paterson GLJ, Coppock R, Sleight V, et al. The deep sea is a major sink for microplastic debris. Royal Society Open Science. 2014;**1**(4):140317. DOI: 10.1098/rsos.140317

[31] Carr SA. Sources and dispersive modes of micro-fibers in the environment. Integrated Environmental Assessment and Management. 2017;**13**(3):466-469. DOI: 10.1002/ieam.1916

[32] Hartline NL, Bruce NJ, Karba SN, Ruff EO, Sonar SU, Holden PA. Microfiber masses recovered from conventional machine washing of new or aged garments. Environmental Science and Technology. 2016;**50**(21):11532-11538. DOI: 10.1021/acs.est.6b03045

[33] Raja Balasaraswathi S, Rathinamoorthy R. Effect of fabric properties on microfiber shedding from synthetic textiles. Journal of the Textile Institute. 2022;**113**(5):789-809. DOI: 10.1080/00405000.2021.1906038

[34] Shafiq A, Johnson F, Klassen RD, Awaysheh A. The impact of supply risk on sustainability monitoring practices and performance. Academy of Management Proceedings. 2016;**2016**(1):17571. DOI: 10.5465/AMBPP.2016.17

[35] de Oliveira CRS, da Silva Júnior AH, Mulinari J, Ferreira AJS, da Silva A. Fibrous microplastics released from textiles: Occurrence, fate, and remediation strategies. Journal of Contaminant Hydrology. 2023;**256**(2022):104169. DOI: 10.1016/j.jconhyd.2023.104169

[36] Palacios-Marín AV, Tausif M. Fragmented fibre (including microplastics) pollution from textiles. Textile Progress. 2021;**53**(3):123-182. DOI: 10.1080/00405167.2022.2066913

[37] Wang M, Yang J, Zheng S, Jia L, Yong ZY, Yong EL, et al. Unveiling the microfiber release footprint: Guiding control strategies in the textile production industry. Environmental Science and Technology. 2023b;**57**(50):21038-21049. DOI: 10.1021/acs.est.3c06210

[38] Lares M, Ncibi MC, Sillanpää M, Sillanpää M. Occurrence, identification and removal of microplastic particles and fibers in conventional activated sludge process and advanced MBR technology.

Water Research. 2018;**133**:236-246. DOI: 10.1016/j.watres.2018.01.049

[39] Magnusson K, Norén F. Screening of microplastic particles in and down-stream a wastewater treatment plant. IVL Swedish Environmental Research Institute. 2014;**C55**(C):22 https://doi.org/naturvardsverket-2226

[40] Murphy F, Ewins C, Carbonnier F, Quinn B. Wastewater treatment works (WwTW) are a source of microplastics in aquatic environments. Environmental Science and Technology. 2016;**50**(11):5800-5808. DOI: 10.1021/acs.est.5b05416

[41] Ziajahromi S, Neale PA, Rintoul L, Leusch FDLL. Wastewater treatment plants as a pathway for microplastics: Development of a new approach to sample wastewater-based microplastics. Water Research. 2017;**112**:93-99. DOI: 10.1016/j.watres.2017.01.042

[42] Browne MA. Sources and pathways of microplastics to habitats. In: Marine Anthropogenic Litter [Internet]. Bremerhaven, Germany: Springer International Publishing; 2015. pp. 229-244. Available from: https://link.springer.com/chapter/10.1007/978-3-319-16510-3_9

[43] Carr SA, Liu J, Tesoro AG. Transport and fate of microplastic particles in wastewater treatment plants. Water Research. 2016;**91**:174-182. DOI: 10.1016/j.watres.2016.01.002

[44] Cole M, Galloway TS. Ingestion of nanoplastics and microplastics by Pacific oyster larvae. Environmental Science and Technology. 2015;**49**(24):14625-14632. DOI: 10.1021/acs.est.5b04099

[45] Cole M, Lindeque P, Fileman E, Halsband C, Goodhead R, Moger J, et al. Microplastic ingestion by zooplankton. Environmental Science and Technology. 2013;**47**(12):6646-6655. DOI: 10.1021/es400663f

[46] Desforges JPW, Galbraith M, Ross PS. Ingestion of microplastics by zooplankton in the Northeast Pacific Ocean. Archives of Environmental Contamination and Toxicology. 2015;**69**(3):320-330. DOI: 10.1007/s00244-015-0172-5

[47] Cole M. A novel method for preparing microplastic fibers. Scientific Reports. 2016;**6**:1-7. DOI: 10.1038/srep34519

[48] Bachmeier CJ, Trickler WJ, Miller DW. Comparison of drug efflux transport kinetics in various blood-brain barrier models. Drug Metabolism and Disposition. 2006;**34**(6):998-1003. DOI: 10.1124/dmd.105.006999

[49] Bakir A, Rowland SJ, Thompson RC. Enhanced desorption of persistent organic pollutants from microplastics under simulated physiological conditions. Environmental Pollution. 2014;**185**:16-23. DOI: 10.1016/j.envpol.2013.10.007

[50] Lusher AL, Burke A, O'Connor I, Officer R. Microplastic pollution in the Northeast Atlantic Ocean: Validated and opportunistic sampling. Marine Pollution Bulletin. 2014;**88**(1-2):325-333. DOI: 10.1016/j.marpolbul.2014.08.023

[51] Lohmann R. Microplastics are not important for the cycling and bioaccumulation of organic pollutants in the oceans—But should microplastics be considered POPs themselves? Integrated Environmental Assessment and Management. 2017;**13**(3):460-465. DOI: 10.1002/ieam.1914

[52] Yang D, Shi H, Li L, Li J, Jabeen K, Kolandhasamy P. Microplastic

pollution in table salts from China. Environmental Science and Technology. 2015;**49**(22):13622-13627. DOI: 10.1021/acs.est.5b03163

[53] Dris R, Gasperi J, Saad M, Mirande C, Tassin B. Synthetic fibers in atmospheric fallout: A source of microplastics in the environment? Marine Pollution Bulletin. 2016;**104**(1-2):290-293. DOI: 10.1016/j.marpolbul.2016.01.006

[54] Kwon S, Venditti RA, Frazier R, Zambrano F, Pawlak JJ, Zambrano MC, et al. Waterborne and airborne microbers shed from non-woven materials in water and air environments. Research Square. 2021:1-23. DOI: 10.21203/rs.3.rs-996055/v1 [Preprint]

[55] Textile Exchange. The Future of Synthetics. USA: Textile Exchange; 2024. Available from: https://textileexchange.org/knowledge-center/reports/the-future-of-synthetics/?gad_source=1&gclid=EAIaIQobChMIwPjdwKqLiQMVo-IWBR1c0w2cEAAYASAAEgK9S_D_BwE

[56] Decitex Smart Textiles. Microfiber. France: Decitex Smart Textiles; 2016. Available from: https://www.decitex.com/en/microfiber#

[57] Roos S, Arturin OL, Hanning A-C. Microplastics Shedding from Polyester Fabrics. Sweden: Mistra Future Fashion; 2017. Available from: https://mistrafuturefashion.mistraprograms.org/publikationer/MFF-Report-Microplastics.pdf

[58] Tao D, Zhang K, Xu S, Lin H, Liu Y, Kang J, et al. Microfibers released into the air from a household tumble dryer. Environmental Science and Technology Letters. 2022;**9**(2):120-126. DOI: 10.1021/acs.estlett.1c00911

[59] Karkkainen N, Sillanpaa M. Quantification of different microplastic fibres discharged from textiles in machine wash and tumble drying. Environmental Science and Pollution Research. 2021;**28**(13):16253-16263. DOI: 10.1007/s11356-020-11988-2

[60] O'Brien S, Okoffo ED, O'Brien JW, Ribeiro F, Wang X, Wright SL, et al. Airborne emissions of microplastic fibres from domestic laundry dryers. Science of the Total Environment. 2020;**747**:141175. DOI: 10.1016/j.scitotenv.2020.141175

[61] Huang YM, He T, Yan MT, Yang L, Gong H, Wang WJ, et al. Atmospheric transport and deposition of microplastics in a subtropical urban environment. Journal of Hazardous Materials. 2021;**416**:126168. DOI: 10.1016/j.jhazmat.2021.126168

[62] Prata JC, Castro JL, da Costa JP, Duarte AC, Rocha-Santos T, Cerqueira M. The importance of contamination control in airborne fibres and microplastic sampling: Experiences from indoor and outdoor air sampling in Aveiro, Portugal. Marine Pollution Bulletin. 2021, 2020;**159**:111888. DOI: 10.1016/j.marpolbul.2020.111888

[63] Wright SL, Ulke J, Font A, Chan KLA, Kelly FJ. Atmospheric microplastic deposition in an urban environment and an evaluation of transport. Environment International. 2020;**136**:105411. DOI: 10.1016/j.envint.2019.105411

[64] Mizraji R, Ahrendt C, Perez-Venegas D, Vargas J, Pulgar J, Aldana M, et al. Is the feeding type related to the microplastic content in the intertidal fish gut? Marine Pollution Bulletin. 2017;**116**(1-2):498-500. DOI: 10.1016/j.marpolbul.2017.01.008

[65] Ragusa A, Svelato A, Santacroce CA, Baiocco F, Draghi S, D'Amore E, et al. Plasticenta: First evidence of

microplastics in the human placenta. Environment International. 2021;**146**:1-2. DOI: 10.1016/j.envint.2020.106274

[66] Amato-Lourenço LF, Carvalho-Oliveira R, Júnior GR, dos Santos Galvão L, Ando RA, Mauad T. Presence of airborne microplastics in human lung tissue. Journal of Hazardous Materials. 2021;**416**(May):126124. DOI: 10.1016/j.jhazmat.2021.126124

[67] Kühn S, Bravo Rebolledo EL, van Franeker JA. Deleterious effects of litter on marine life. In: Marine Anthropogenic Litter [Internet]. Bremerhaven, Germany: Springer; 2015. pp. 75-116. DOI: 10.1007/978-3-319-16510-3_4

[68] Sundt P, Schultze P-E, Syversen F. Sources of Microplastic- Pollution to the Marine Environment. Norway: Mepex, Norwegian Environment Agency (Issue January); 2014. Available from: https://www.miljodirektoratet.no/globalassets/publikasjoner/M321/M321.pdf

[69] Miller RZ, Watts AJR, Winslow BO, Galloway TS, Barrows APW. Mountains to the sea: A river study of plastic and non-plastic microfiber pollution in Northeast USA. Marine Pollution Bulletin. 2017;**124**(1):245-251. DOI: 10.1016/j.marpolbul.2017.07.028

[70] Dris R, Gasperi J, Rocher V, Tassin B. Synthetic and non-synthetic anthropogenic fibers in a river under the impact of Paris megacity: Sampling methodological aspects and flux estimations. Science of the Total Environment. 2018;**618**:157-164. DOI: 10.1016/j.scitotenv.2017.11.009

[71] Dris R, Gasperi J, Tassin B. Sources and fate of microplastics in urban areas: A focus on Paris megacity. In: Handbook of Environmental Chemistry. Vol. 58. Cham: Springer; 2018b. pp. 69-83. DOI: 10.1007/978-3-319-61615-5_4

[72] Julapong P, Srichonphaisarn P, Meekoch T, Tabelin CB, Juntarasakul O, Phengsaart T. The influence of textile type, textile weight, and detergent dosage on microfiber emissions from top-loading washing machines. Toxics. 2024;**12**(3):210. DOI: 10.3390/toxics12030210

[73] Periyasamy AP. Evaluation of microfiber release from jeans: Impact of different washing conditions. Environmental Science and Pollution Research. 2021;**28**(41):58570-58582. DOI: 10.1007/s11356-021-14761-1

[74] Wang C, Chen W, Zhao H, Tang J, Li G, Zhou Q, et al. Microplastic fiber release by laundry: A comparative study of hand-washing and machine-washing. ACS ES and T Water. 2023a;**3**(1):147-155. DOI: 10.1021/acsestwater.2c00462

[75] Pirc U, Vidmar M, Mozer A, Kržan A. Emissions of microplastic fibres from microfiber fleece during domestic washing. Environmental Science and Pollution Research. 2016;**23**(21):22206-22211. DOI: 10.1007/s11356-016-7703-0

[76] Sillanpää M, Sainio P. Release of polyester and cotton fibers from textiles in machine washings. Environmental Science and Pollution Research. 2017;**24**(23):19313-19321. DOI: 10.1007/s11356-017-9621-1

[77] Kelly MR, Lant NJ, Kurr M, Burgess JG. Importance of water-volume on the release of microplastic fibers from laundry. Environmental Science and Technology. 2019;**53**(20):11735-11744. DOI: 10.1021/acs.est.9b03022

[78] Cesa FSFSFS, Turra A, Checon HHHHHH, Leonardi B, Baruque-Ramos J, Leonardi B, et al.

Laundering and textile parameters influence fibers release in household washings. Environmental Pollution. 2020;**257**:113553. DOI: 10.1016/j.envpol.2019.113553

[79] Choi S, Kwon M, Park MJ, Kim J. Characterization of microplastics released based on polyester fabric construction during washing and drying. Polymers. 2021;**13**(24):4277. DOI: 10.3390/polym13244277

[80] Palacios-Marín AV, Jabbar A, Tausif M. Fragmented fibre pollution from common textile materials and structures during laundry. Textile Research Journal. 2022;**92**(13-14):2265-2275. DOI: 10.1177/00405175221090971

[81] Mahbub MS, Shams M. Acrylic fabrics as a source of microplastics from portable washer and dryer: Impact of washing and drying parameters. Science of the Total Environment. 2022;**834**(February):155429. DOI: 10.1016/j.scitotenv.2022.155429

[82] Hazlehurst A, Tiffin L, Sumner M, Taylor M. Quantification of microfibre release from textiles during domestic laundering. Environmental Science and Pollution Research. 2023;**30**(15):43932-43949. DOI: 10.1007/s11356-023-25246-8

[83] Volgare M, De Falco F, Avolio R, Castaldo R, Errico ME, Gentile G, et al. Washing load influences the microplastic release from polyester fabrics by affecting wettability and mechanical stress. Scientific Reports. 2021;**11**(1):19479-19479. DOI: 10.1038/s41598-021-98836-6

[84] Lant NJ, Hayward AS, Peththawadu MMD, Sheridan KJ, Dean JR. Microfiber release from real soiled consumer laundry and the impact of fabric care products and washing conditions. PLoS One. 2020;**15**(6):e0233332-e0233332. DOI: 10.1371/journal.pone.0233332

[85] Hernandez E, Nowack B, Mitrano DM. Polyester textiles as a source of microplastics from households: A mechanistic study to understand microfiber release during washing. Environmental Science and Technology. 2017;**51**(12):7036-7046. DOI: 10.1021/acs.est.7b01750

[86] De Falco F, Gullo MP, Gentile G, Di Pace E, Cocca M, Gelabert L, et al. Evaluation of microplastic release caused by textile washing processes of synthetic fabrics. Environmental Pollution. 2018b;**236**:916-925. DOI: 10.1016/j.envpol.2017.10.057

[87] Zambrano MC, Pawlak JJ, Daystar J, Ankeny M, Cheng JJ, Venditti RA. Microfibers generated from the laundering of cotton, rayon and polyester based fabrics and their aquatic biodegradation. Marine Pollution Bulletin. 2019;**142**(February):394-407. DOI: 10.1016/j.marpolbul.2019.02.062

[88] Yang L, Qiao F, Lei K, Li H, Kang Y, Cui S, et al. Microfiber release from different fabrics during washing. Environmental Pollution. 2019b;**249**:136-143. DOI: 10.1016/j.envpol.2019.03.011

[89] Cotton L, Hayward AS, Lant NJ, Blackburn RS. Improved garment longevity and reduced microfibre release are important sustainability benefits of laundering in colder and quicker washing machine cycles. Dyes and Pigments. 2020;**177**:108120. DOI: 10.1016/j.dyepig.2019.108120

[90] McIlwraith HK, Lin J, Erdle LM, Mallos N, Diamond ML, Rochman CM. Capturing microfibers – Marketed technologies reduce microfiber emissions from washing machines. Marine

Pollution Bulletin. 2019;**139**:40-45. DOI: 10.1016/j.marpolbul.2018.12.012

[91] Napper IE, Barrett AC, Thompson RC. The efficiency of the devices is intended to reduce microfiber release during washing clothes. Science of the Total Environment. 2020;**738**:140412. DOI: 10.1016/j.scitotenv.2020.140412

[92] Herweyers L, Carteny CC, Scheelen L, Watts R, Bois ED. Consumers' perceptions and attitudes towards products preventing microfiber pollution in aquatic environments as a result of the domestic washing of synthetic clothes. Sustainability (Switzerland). 2020;**12**(6):12-14. DOI: 10.3390/su12062244

[93] Özkan İ, Gündoğdu S. Investigation of microfiber release under controlled washing from knitted fabrics produced by recycled and virgin polyester yarns. Journal of the Textile Institute. 2021;**112**(2):264-272. DOI: 10.1080/00405000.2020.1741760

[94] De Falco F, Di Pace E, Cocca M. MCM, Avella M, Falco F. De, Pace E. Di, Cocca MMCM, Avella M, F Falco. The contribution of washing processes of synthetic clothes to microplastic pollution. Scientific Reports. 2019;**9**(1):6633. doi:10.1038/s41598-019-43023-x

[95] Celik S. Microplastic release from domestic washing. European Journal of Science and Technology. 2021;**25**:790-795. DOI: 10.31590/ejosat.933322

[96] Rathinamoorthy R, Raja Balasaraswathi S, Madhubashini S, Prakalya A, Rakshana JB, Shathvika S. Investigation on microfiber release from elastane blended fabrics and its environmental significance. Science of the Total Environment. 2023;**903**(June):166553. DOI: 10.1016/j.scitotenv.2023.166553

[97] Belzagui F, Crespi M, Álvarez A, Gutiérrez-Bouzán C, Vilaseca M. Microplastics' emissions: Microfibers' detachment from textile garments. Environmental Pollution. 2019;**248**:1028-1035. DOI: 10.1016/j.envpol.2019.02.059

[98] De Falco F, Cocca M, Avella M, Thompson RC, Falco F. De, Cocca M, Avella M, Thompson RC. Microfiber release to water, via laundering, and to air, via everyday use: A comparison between polyester clothing with differing textile parameters. Environmental Science and Technology. 2020;**54**(6):3288-3296. DOI: 10.1021/acs.est.9b06892

[99] Cai Y, Mitrano DM, Heuberger M, Hufenus R, Nowack B. The origin of microplastic fiber in polyester textiles: The textile production process matters. Journal of Cleaner Production. 2020a;**267**:121970. DOI: 10.1016/j.jclepro.2020.121970

[100] Hazlehurst A, Sumner M, Taylor M. Investigating the influence of yarn characteristics on microfibre release from knitted fabrics during laundering. Frontiers in Environmental Science. 2024;**12**(March):1-13. DOI: 10.3389/fenvs.2024.1340229

[101] Carney Almroth BM, Åström L, Roslund S, Petersson H, Johansson M, Persson NK. Quantifying shedding of synthetic fibers from textiles; a source of microplastics released into the environment. Environmental Science and Pollution Research. 2018;**25**(2):1191-1199. DOI: 10.1007/s11356-017-0528-7

[102] Vassilenko E, Watkins M, Chastain S, Mertens J, Posacka AM, Patankar S, et al. Domestic laundry and microfiber pollution: Exploring fiber

shedding from consumer apparel textiles. PLoS One. 2021;**16**(7 July):e0250346. DOI: 10.1371/journal.pone.0250346

[103] Berruezo M, Bonet-Aracil M, Montava I, Bou-Belda E, Díaz-García P, Gisbert-Payá J. Preliminary study of weave pattern influence on microplastics from fabric laundering. Textile Research Journal. 2021;**91**(9-10):1037-1045. DOI: 10.1177/0040517520965708

[104] Rathinamoorthy R, Raja Balasaraswathi S. Investigations on the impact of handwashing and laundry softeners on microfiber shedding from polyester textiles. Journal of the Textile Institute. 2022;**113**(7):1428-1437. DOI: 10.1080/00405000.2021.1929709

[105] De Falco F, Gentile G, Avolio R, Errico ME, Di Pace E, Ambrogi V, et al. Pectin based finishing to mitigate the impact of microplastics released by polyamide fabrics. Carbohydrate Polymers. 2018a;**198**(March):175-180. DOI: 10.1016/j.carbpol.2018.06.062

[106] Cui H, Xu C. Study on the relationship between textile microplastics shedding and fabric structure. Polymers. 2022;**14**(23). DOI: 10.3390/polym14235309

[107] Klinkhammer K, Kolbe S, Brandt S, Meyer J, Ratovo K, Bendt E, et al. Release of fibrous microplastics from functional polyester garments through household washing. Frontiers in Environmental Science. 2024;**12**(April):1-15. DOI: 10.3389/fenvs.2024.1330922

[108] Cai Y, Yang T, Mitrano DM, Heuberger M, Hufenus R, Nowack B. Systematic study of microplastic fiber release from 12 different polyester textiles during washing. Environmental Science and Technology. 2020b;**54**(8): 4847-4855. DOI: 10.1021/acs.est.9b07395

[109] Choi S, Kim J, Kwon M. The effect of the physical and chemical properties of synthetic fabrics on the release of microplastics during washing and drying. Polymers. 2022;**14**(16). DOI: 10.3390/polym14163384

[110] Dalla Fontana G, Mossotti R, Montarsolo A. Influence of sewing on microplastic release from textiles during washing. Water, Air, and Soil Pollution. 2021;**232**(2):1-9. DOI: 10.1007/s11270-021-04995-7

[111] Rathinamoorthy R, Raja Balasaraswathi S. Impact of quantification method on microfiber assessment: A comparative analysis between mass- and count-based methods. Journal of Environmental Management. 2023;**347**(July):119040. DOI: 10.1016/j.jenvman.2023.119040

[112] Rathinamoorthy R, Raja Balasaraswathi S. Assessment of the contribution of sewing threads to microfiber release during domestic laundering. Environmental Pollution. 2024;**362**(June):124966. DOI: 10.1016/j.envpol.2024.124966

[113] Brodin M, Norin H, Hanning A.-C, Persson C. Filters for Washing Machines Mitigation of Microplastic Pollution. 25. 2018a. Available from: http://www.naturvardsverket.se/Documents/publ-kompl/1003-09ReportFilters forwashing machines.pdf

[114] Brodin M, Norin H, Hanning AC, Persson C, Okcabol S. Microplastics from Industrial Laundries - a Study of Laundry Effluents. Sweden: The Swedish Environmental Protection Agency Report; 2018b. Available from: http://www.diva-portal.org/smash/get/diva2:1633776/FULLTEXT01.pdf

[115] Ryan PG, Suaria G, Perold V, Pierucci A, Bornman TG,

Aliani S. Sampling microfibres at the sea surface: The effects of mesh size, sample volume and water depth. Environmental Pollution. 2020;**258**:113413. DOI: 10.1016/j.envpol.2019.113413

[116] Hidalgo-Ruz V, Gutow L, Thompson RC, Thiel M. Microplastics in the marine environment: A review of the methods used for identification and quantification. Environmental Science and Technology. 2012;**46**:3060-3075. DOI: 10.1021/es2031505

[117] Ballent A, Corcoran PL, Madden O, Helm PA, Longstaffe FJ. Sources and sinks of microplastics in Canadian Lake Ontario nearshore, tributary and beach sediments. Marine Pollution Bulletin. 2016;**110**(1):383-395. DOI: 10.1016/j.marpolbul.2016.06.037

[118] The Microfibre Consortium. The TMC Test Method [Internet]. United Kingdom: The Microfibre Consortium; 2021. Available from: https://www.microfibreconsortium.com/the-tmc-test-method

[119] AATCC. AATCC TM212-2021 Test Method for Fiber Fragment Release During Home Laundering. AATCC; 2021. Available from: https://members.aatcc.org/store/tm212/3573/

[120] ISO. ISO - ISO/DIS 4484-1 - Textiles and Textile Products — Microplastics from Textile Sources — Part 1: Determination of Material Loss from Fabrics during Washing. Geneva, Switzerland: ISO; 2022a. Available from: https://www.iso.org/standard/80011.html

[121] The Microfibre Consortium. TMC Progress Update Report. United Kingdom: The Microfibre Consortium; 2024. Available from: https://www.microfibreconsortium.com/tmc-progress-report-22-24

[122] Sun J, Dai X, Wang Q, van Loosdrecht MCM, Ni BJ. Microplastics in wastewater treatment plants: Detection, occurrence and removal. Water Research. 2019;**152**:21-37. DOI: 10.1016/j.watres.2018.12.050

[123] Mintenig SM, Int-Veen I, Löder MGJ, Primpke S, Gerdts G. Identification of microplastics in wastewater treatment plant effluents using focal plane array-based micro-Fourier-transform infrared imaging. Water Research. 2017;**108**(1):365-372. DOI: 10.1016/j.watres.2016.11.015

[124] Conley K, Clum A, Deepe J, Lane H, Beckingham B. Wastewater treatment plants as a source of microplastics to an urban estuary: Removal efficiencies and loading per capita over one year. Water Research X. 2019;**3**:100030. DOI: 10.1016/j.wroa.2019.100030

[125] Talvitie J, Mikola A, Koistinen A, Setälä O. Solutions to microplastic pollution – Removal of microplastics from wastewater effluent with advanced wastewater treatment technologies. Water Research. 2017;**123**:401-407. DOI: 10.1016/j.watres.2017.07.005

[126] Talvitie J, Mikola A, Setälä O, Heinonen M, Koistinen A. How well is microlitter purified from wastewater? – A detailed study on the stepwise removal of microlitter in a tertiary level wastewater treatment plant. Water Research. 2017b;**109**:164-172. DOI: 10.1016/j.watres.2016.11.046

[127] Xu X, Hou Q, Xue Y, Jian Y, Wang LP. Pollution characteristics and fate of microfibers in the wastewater from textile dyeing wastewater treatment plant. Water Science and Technology. 2018;**78**(10):2046-2054. DOI: 10.2166/wst.2018.476

[128] Yang L, Li K, Cui S, Kang Y, An L, Lei K. Removal of microplastics in

municipal sewage from China's largest water reclamation plant. Water Research. 2019a;**155**:175-181. DOI: 10.1016/j.watres.2019.02.046

[129] Lv X, Dong Q, Zuo Z, Liu Y, Huang X, Wu WM. Microplastics in a municipal wastewater treatment plant: Fate, dynamic distribution, removal efficiencies, and control strategies. Journal of Cleaner Production. 2019;**225**:579-586. DOI: 10.1016/j.jclepro.2019.03.321

[130] Rather LJ, Jameel S, Dar OA, Ganie SA, Bhat KA, Mohammad F. Advances in sustainable technologies for water conservation in the textile industry. In: Water in Textiles and Fashion. United Kingdom: Woodhead Publishing; 2019. pp. 175-194. DOI: 10.1016/b978-0-08-102633-5.00010-5

[131] Habib D, Locke DC, Cannone LJ. Synthetic fibers as indicators of municipal sewage. Water, Air and Soil Pollution. 1998;**103**(103):1-8. Available from: https://link.springer.com/content/pdf/10.1023%2FA%3A1004908110793.pdf

[132] Zubris KAV, Richards BK. Synthetic fibers as an indicator of land application of sludge. Environmental Pollution. 2005;**138**(2):201-211. DOI: 10.1016/j.envpol.2005.04.013

[133] Gago J, Carretero O, Filgueiras AV, Viñas L. Synthetic microfibers in the marine environment: A review on their occurrence in seawater and sediments. Marine Pollution Bulletin. 2018;**127**(2017):365-376. DOI: 10.1016/j.marpolbul.2017.11.070

[134] Baldwin AK, Corsi SR, Mason SA. Plastic debris in 29 Great Lakes tributaries: Relations to watershed attributes and hydrology. Environmental Science and Technology. 2016;**50**(19):10377-10385. DOI: 10.1021/acs.est.6b02917

[135] Klein S, Worch E, Knepper TP. Occurrence and spatial distribution of microplastics in river shore sediments of the Rhine-Main Area in Germany. Environmental Science & Technology. 2015;**49**(10):6070-6076. DOI: 10.1021/acs.est.5b00492

[136] Sanchez-Vidal A, Thompson RC, Canals M, de Haan WP. The imprint of microfibres in southern European deep seas. Chin WC, editor. PLoS One. 2018;**13**(11):e0207033. DOI: 10.1371/journal.pone.0207033

[137] Vermaire JC, Pomeroy C, Herczegh SM, Haggart O, Murphy M. Microplastic abundance and distribution in the open water and sediment of the Ottawa River, Canada, and its tributaries. FACETS. 2017;**2**(1):301-314. DOI: 10.1139/facets-2016-0070

[138] Cole M, Lindeque P, Halsband C, Galloway TS. Microplastics as contaminants in the marine environment: A review. Marine Pollution Bulletin. 2011;**62**(12):2588-2597. DOI: 10.1016/J.MARPOLBUL.2011.09.025

[139] Dris R, Gasperi J, Mirande C, Mandin C, Guerrouache M, Langlois V, et al. A first overview of textile fibers, including microplastics, in indoor and outdoor environments. Environmental Pollution. 2017;**221**:453-458. DOI: 10.1016/j.envpol.2016.12.013

[140] Liu K, Wang XH, Fang T, Xu P, Zhu LX, Li DJ. Source and potential risk assessment of suspended atmospheric microplastics in Shanghai. Science of the Total Environment. 2019;**675**:462-471. DOI: 10.1016/j.scitotenv.2019.04.110

[141] Zhang Q, Zhao Y, Du F, Cai H, Wang G, Shi H. Microplastic fallout

in different indoor environments. Environmental Science and Technology. 2020;**54**(11):6530-6539. DOI: 10.1021/acs.est.0c00087

[142] Sharaf Din K, Khokhar MF, Parvez S, Niazi MBK. Emerging environmental challenges: A critical review of airborne microplastics. Environmental Research Communications. 2024;**6**. DOI: 10.1088/2515-7620/ad75eb

[143] Abbasi S, Turner A. Dry and wet deposition of microplastics in a semi-arid region (Shiraz, Iran). Science of the Total Environment. 2021;**786**:147358. DOI: 10.1016/j.scitotenv.2021.147358

[144] Prata JC, Castro JL, da Costa JP, Duarte AC, Rocha-Santos T, Cerqueira M. Importance of contamination control in airborne fibres and microplastic sampling: Experiences from indoor and outdoor air sampling in Aveiro, Portugal. Marine Pollution Bulletin. 2020;**159**(August):111522. DOI: 10.1016/j.marpolbul.2020.111522

[145] Torres-Agullo A, Karanasiou A, Moreno T, Lacorte S. Airborne microplastic particle concentrations and characterization in indoor urban microenvironments. Environmental Pollution. 2022;**308**(March):119707. DOI: 10.1016/j.envpol.2022.119707

[146] Gaston E, Woo M, Steele C, Sukumaran S, Anderson S. Microplastics differ between indoor and outdoor air masses: Insights from multiple microscopy methodologies. Applied Spectroscopy. 2020;**74**(9):1079-1098. DOI: 10.1177/0003702820920652/FORMAT/EPUB

[147] Tunahan Kaya A, Yurtsever M, Çiftçi Bayraktar S. Ubiquitous exposure to microfiber pollution in the air. European Physical Journal Plus. 2018;**133**(11):488. DOI: 10.1140/epjp/i2018-12372-7

[148] Taylor ML, Gwinnett C, Robinson LF, Woodall LC. Plastic microfibre ingestion by deep-sea organisms. Scientific Reports. 2016;**6**(1):33997. DOI: 10.1038/srep33997

[149] Mathalon A, Hill P. Microplastic fibers in the intertidal ecosystem surrounding Halifax Harbor, Nova Scotia. Marine Pollution Bulletin. 2014;**81**(1):69-79. DOI: 10.1016/j.marpolbul.2014.02.018

[150] Collard F, Gilbert B, Eppe G, Parmentier E, Das K. Detection of anthropogenic particles in fish stomachs: An isolation method adapted to identification by Raman spectroscopy. Archives of Environmental Contamination and Toxicology. 2015;**69**(3):331-339. DOI: 10.1007/s00244-015-0221-0

[151] Keshavarzifard M, Zakaria MP, Sharifi R. Ecotoxicological and health risk assessment of polycyclic aromatic hydrocarbons (PAHs) in short-neck clam (Paphia undulata) and contaminated sediments in Malacca Strait, Malaysia. Archives of Environmental Contamination and Toxicology. 2017;**73**(3):474-487. DOI: 10.1007/s00244-017-0410-0

[152] Watts AJR, Urbina MA, Corr S, Lewis C, Galloway TS. Ingestion of plastic microfibers by the crab *Carcinus maenas* and its effect on food consumption and energy balance. Environmental Science and Technology. 2015;**49**(24):14597-14604. DOI: 10.1021/acs.est.5b04026

[153] Lehtiniemi M, Hartikainen S, Näkki P, Engström-öst J, Koistinen A, Setälä O. Size matters more than shape: Ingestion of primary and secondary

microplastics by small predators. Food Webs. 2018;**17**:e00097. DOI: 10.1016/j.fooweb.2018.e00097

[154] Athey SN, Erdle LM. Are we underestimating anthropogenic microfiber pollution? A critical review of occurrence, methods, and reporting. Environmental Toxicology and Chemistry. 2022;**41**(4):822-837. DOI: 10.1002/etc.5173

[155] Wright SL, Thompson RC, Galloway TS. The physical impacts of microplastics on marine organisms: A review. Environmental Pollution (Barking, Essex: 1987). 2013;**178**:483-492. DOI: 10.1016/j.envpol.2013.02.031

[156] Galloway TS. Micro- and nanoplastics and human health. In: Marine Anthropogenic Litter. Bremerhaven, Germany: Springer International Publishing; 2015. pp. 343-366. DOI: 10.1007/978-3-319-16510-3_13

[157] Erni-Cassola G, Gibson MI, Thompson RC, Christie-Oleza JA. Lost, but found with Nile red: A novel method for detecting and quantifying small microplastics (1 mm to 20 μm) in environmental samples. Environmental Science and Technology. 2017;**51**(23):13641-13648. DOI: 10.1021/acs.est.7b04512

[158] Gasperi J, Wright SL, Dris R, Collard F, Mandin C, Guerrouache M, et al. Microplastics in air: Are we breathing it in? Current Opinion in Environmental Science and Health. 2018;**1**:1-5. DOI: 10.1016/j.coesh.2017.10.002

[159] Jenner LC, Rotchell JM, Bennett RT, Cowen M, Tentzeris V, Sadofsky LR. Detection of microplastics in human lung tissue using μFTIR spectroscopy. Science of the Total Environment. 2022;**831**(March):154907. DOI: 10.1016/j.scitotenv.2022.154907

[160] Luqman A, Nugrahapraja H, Wahyuono RA, Islami I, Haekal MH, Fardiansyah Y, et al. Microplastic contamination in human stools, foods, and drinking water associated with Indonesian coastal population. Environments. 2021;**8**(12):1-9. DOI: 10.3390/environments8120138

[161] Schwabl P, Koppel S, Konigshofer P, Bucsics T, Trauner M, Reiberger T, et al. Detection of Various Microplastics in Human Stool: A Prospective Case Series. USA: Annals of Internal Medicine, American College of Physicians; 2019. DOI: 10.7326/M19-0618

[162] Xia X, Sun M, Zhou M, Chang Z, Li L. Polyvinyl chloride microplastics induce growth inhibition and oxidative stress in *Cyprinus carpio* var. larvae. Science of the Total Environment. 2020;**716**. DOI: 10.1016/j.scitotenv.2019.136479

[163] Wu D, Feng Y, Wang R, Jiang J, Guan Q, Yang X, et al. Pigment microparticles and microplastics found in human thrombi based on Raman spectral evidence. Journal of Advanced Research. 2023;**49**:141-150. DOI: 10.1016/j.jare.2022.09.004

[164] Emenike EC, Okorie CJ, Ojeyemi T, Egbemhenghe A, Iwuozor KO, Saliu OD, et al. From oceans to dinner plates: The impact of microplastics on human health. Heliyon. 2023;**9**(10):e20440. DOI: 10.1016/j.heliyon.2023.e20440

[165] Yan S, Jones C, Henninger CE, McCormick H. Textile industry insights towards impact of regenerated cellulosic and synthetic fibres on microfibre pollution. Sustainable Textiles: Production, Processing, Manufacturing & Chemistry. 2020:157-171. DOI: 10.1007/978-3-030-38013-7_8

[166] Yu Y, Chen H, Hua X, Dang Y, Han Y, Yu Z, et al. Polystyrene microplastics (PS-MPs) toxicity induced oxidative stress and intestinal injury in nematode Caenorhabditis elegans. Science of the Total Environment. 2020;**726**:138679. DOI: 10.1016/j.scitotenv.2020.138679

[167] Chan CKM, Fang JKH, Fei B, Kan CW. Microfibres release from textile industry wastewater effluents are underestimated: Mitigation actions that need to be prioritised. Fibers. 2023;**11**(12):1-27. DOI: 10.3390/fib11120105

[168] Kapp KJ, Miller RZ. Electric clothes dryers: An underestimated source of microfiber pollution. PLoS One. 2020;**15**(10 October):1-17. DOI: 10.1371/journal.pone.0239165

[169] Xu C, Zhang B, Gu C, Shen C, Yin S, Aamir M, et al. Are we underestimating the sources of microplastic pollution in terrestrial environment? Journal of Hazardous Materials. 2020;**400**:123228. DOI: 10.1016/j.jhazmat.2020.123228

[170] Dreillard M, Barros CDF, Rouchon V, Emonnot C, Lefebvre V, Moreaud M, et al. Quantification and morphological characterization of microfibers emitted from textile washing. Science of the Total Environment. 2022;**832**(January):154973. DOI: 10.1016/j.scitotenv.2022.154973

[171] Chan CKM, Curie P, Ming CK, Park C, Chan KM, Mak DCW, et al. Microplastic fibre releases from industrial wastewater effluent: A textile wet-processing mill in China. Environmental Chemistry. 2021;**18**(3):93-100. DOI: 10.1071/EN20143

[172] Frost H, Zambrano MC, Leonas K, Pawlak JJ, Venditti RA. Do recycled cotton or polyester fibers influence the shedding propensity of fabrics during laundering? AATCC Journal of Research. 2020;**7**(Special Issue 1):32-41. DOI: 10.14504/ajr.7.S1.4

[173] Wang K, Wang L, Xiangyu Y, Wu M, Liu H. Effects and characterization of environmental conditions on microplastic fibers release from synthetic textile. Environmental Science and Pollution Research. 2024:1-18. DOI: 10.21203/RS.3.RS-3758709/V1 [Preprint]

[174] Plastic Soup Foundation. Microfiber Pollution: What Are the Solutions for the Oceans? Amsterdam, Netherlands: Plastic Soup Foundation; 2022. Available from: https://www.oceancleanwash.org/solutions/solutions-for-consumers/

[175] Changing Markets Foundation. Fashion's Plastic Paralysis: How Brands Resist Change and Fuel Microplastic Pollution. Utrecht, Netherlands: Changing Markets Foundation; 2024. Available from: https://changingmarkets.org/report/fashions-plastic-paralysis/

[176] ISO 4484-2:2023 [Internet]. ISO. Available from: https://www.iso.org/standard/80011.html

[177] The Microfibre Consortium. TMC Method FAQs. No. Feb, 2019. The Microfibre Consortium Leading the Textile Industry in Reducing Microfibre Release to Protect Our Environment 1 FAQs on the TMC Test Method [Internet]. United Kingdom; 2019. Available from: https://static1.squarespace.com/static/5aaba1998f513028aeec604c/t/611bd9da5a5b2e4c66623c21/1629215194569/TMC+Test+Method_FAQs.pdf [Accessed: December 20, 2024]

[178] Kentin E, Battaglia G. Policies and perspectives on regulating microplastic fibre pollution. In: Polluting Textiles: The Problem with Microfibres.

Chapter 13. 2022. pp. 265-289. DOI: 10.4324/9781003165385-13

[179] Li Y. Legislation and policy on pollution prevention and the control of marine microplastics. Water (Switzerland). 2022;**14**(18). DOI: 10.3390/w14182790

[180] European Parliament. Making it Mandatory at EU Level to Fit Washing Machines with a Microplastic Filter E-002525/2023. European Union: European Parliament; 2023. Available from: https://www.europarl.europa.eu/doceo/document/E-9-2023-002525_EN.html?form=MG0AV3

[181] Sancroft. Microfibres Pollution: Current State Assessment Report. United Kingdom: Sancroft; 2018

[182] Chauhan NS, Punia A. Role of education and society in dealing plastic pollution in the future. In: Plastic and Microplastic in the Environment: Management and Health Risks. Chapter 14. 2022. pp. 276-281. DOI: 10.1002/9781119800897.CH14

[183] Auta HSS, Emenike CU, Fauziah SH. Distribution and importance of microplastics in the marine environment: A review of the sources, fate, effects, and potential solutions. Environment International. 2017;**102**:165-176. DOI: 10.1016/j.envint.2017.02.013

[184] Asmita K, Shubhamsingh T, Tejashree S. Isolation of plastic-degrading microorganisms from soil samples collected at various locations in Mumbai, India. Environmental Science. 2015;**4**(3):77-85. Available from: https://pdfs.semanticscholar.org/5f34/8903e9b2eef3588bd3753692d3ff949a3210.pdf

[185] Ocean Cleaning Device Succeeds in Removing Plastic for the First Time [Internet]. Ohio, USA: EcoWatch; 2019. Available from: https://www.ecowatch.com/ocean-cleaning-device-plastic-2640832397.html

[186] Mermaids Consortium. Microfiber Release from Clothes after Washing: Hard Facts, Figures and Promising Solutions. Amsterdam, Netherlands: Mermaids Consortium; May 2017. pp. 1-9. Position Paper, Available from: https://www.plasticsoupfoundation.org/wp-content/uploads/2017/08/Position-Paper.Microfiber-release-from-clothes-after-washing.PSF_.pdf

[187] PANGAIA x MTIX Microfiber Mitigation — Microfiber Innovation Challenge. Microfiber Innovation Challenge [Internet]. Microfiber Innovation Challenge; 2021. Available from: https://www.microfiberinnovation.org/innovation/pangaia-x-mtix

[188] MTIX LTD - Innovation in Textile Processing Technology [Internet]. mti-x.com. Available from: https://mti-x.com/

[189] FAQ|Mango Materials [Internet]. Mango Materials. 2017. Available from: https://www.mangomaterials.com/faq/

[190] Tandem Repeat. Technology – Tandem Repeat [Internet]. n.d. Available from: https://www.tandemrepeat.com/technology/

[191] Natural Fibre Welding Inc. Natural Fiber Welding, Inc. [Internet]. www.naturalfiberwelding.com. 2024. n.d. Available from: https://www.naturalfiberwelding.com

[192] Yousef S, Eimontas J, Zakarauskas K, Striūgas N, Mohamed A. A new strategy for using lint-microfibers generated from clothes dryer as a sustainable source of renewable energy. Science of the Total

Environment. 2021;**762**. DOI: 10.1016/j.scitotenv.2020.143107

[193] Xu Z, Bai X, Ye Z. Removal and generation of microplastics in wastewater treatment plants: A review. Journal of Cleaner Production. 2021b;**291**:125982. DOI: 10.1016/j.jclepro.2021.125982

[194] Zhou H, Zhou L, Ma K. Microfiber from textile dyeing and printing wastewater of a typical industrial park in China: Occurrence, removal and release. Science of the Total Environment. 2020;**739**:140329. DOI: 10.1016/j.scitotenv.2020.140329

[195] Westphalen H, Abdelrasoul A. Challenges and treatment of microplastics in water. In: Water Challenges of an Urbanizing World. United Kingdom: IntechOpen; 2018. DOI: 10.5772/intechopen.71494

[196] Gavigan J, Kefela T, Macadam-Somer I, Suh S, Geyer R. Synthetic microfiber emissions to land rival those to waterbodies and are growing. PLoS One. 2020;**15**(9 September):1-13. DOI: 10.1371/journal.pone.0237839

[197] Iqbal SA, Mahmud I, Quader AKMAMA. Textile sludge management by incineration technique. Procedia Engineering. 2014;**90**:686-691. DOI: 10.1016/j.proeng.2014.11.795

[198] Ramasamy R, Subramanian RB. Microfiber mitigation from synthetic textiles: Impact of combined surface modification and finishing process. Environmental Science and Pollution Research. 2023;**30**(17):49136-49149

[199] Cowger W, Booth AM, Hamilton BM, Thaysen C, Primpke S, Munno K, et al. Reporting guidelines to increase the reproducibility and comparability of research on microplastics. Applied Spectroscopy. 2020;**74**(9):1066-1077. DOI: 10.1177/0003702820930292

[200] ZWE. A New Look for the Fashion Industry - EU Textile Strategy and the Crucial Role of Extended Producer Responsibility. USA: Changing Markets Foundation; 2022. Available from: https://changingmarkets.org/report/new-look-for-the-fashion-industry-eu-textile-strategy-crucial-role-extended-producer-responsibility-epr/

[201] Harsha V, Abhiram P, Prakash DB. Theoretical review on microplastic pollution: A multifaceted threat to marine ecosystems, human health, and environment. IgMin Research. 2024;**2**(6):460-468. DOI: 10.61927/igmin203

[202] Nafea TH, Ka F, Xu H, Wang C, Xiao H, He J. Status of management and mitigation of microplastic pollution. Critical Reviews in Environmental Science and Technology. 2024;**54**(24):1-23. DOI: 10.1080/10643389.2024.2361502

[203] Allen E, Henninger CE, Garforth A, Asuquo E. Microfiber pollution: A systematic literature review to overcome the complexities in knit design to create solutions for knit fabrics. Environmental Science and Technology. 2024;**58**(9):4031-4045. DOI: 10.1021/acs.est.3c05955